CALIFORNIA GOLD AND THE HIGHGRADERS

True Stories of the Mines and the Miners

To Jim and Betty.

F.D. Calhoon

F.D. CALHOON

CAL-CON PRESS

3501 Bradshaw Rd. #98
Sacramento, CA 95827

ISBN 945862-00-8

DEDICATION

To Mike McGuire whose memory inspired the recording of the stories he told of his mining experiences and his association with the Highgraders.

ACKNOWLEDGEMENTS

To all those who, for more than a hundred years, risked the dangers of working in the deep mines of the Sierra, and lived to tell the stories of their adventures.

To the following for allowing the use of their records and their photographs:

The California State Bureau of Mines

The California State Library

The Grass Valley Union

The Nevada County Union

The California Department of Parks and Recreation

The Park Rangers at the Empire State Park

Gold and the Highgraders

TABLE OF CONTENTS

Gold and the Highgraders

CHAPTER

CHAPTER I

THE FIRST HIGHGRADE

Highgrading: "The taking of gold from another man's property," without permission or payment, first happened in California, by, of all people, The Mormons!

The gold in question belonged to Captain John Sutter. In 1847, he purchased several thousand acres of timbered land on the banks of the Middle Fork of the American River. He needed lumber to carry on his building projects in and near his fort at the junction of the American and the Sacramento Rivers.

At the beginning of the Mexican War several hundred young Mormons enlisted in the United States Army with the promise that they would be sent to California. They, upon their discharge, would be on hand to help Brigham Young begin a settlement near what is now the city of Stockton.

The Mexican War was over before the Mormons arrived. Young sent word for them to march to the Great Salt Lake. He had decided to found his settlement there.

Snow blocked the mountain passes. Sutter needed workmen to build his sawmill. Some twenty Mormons agreed to furnish the skilled labor. In January, 1848, James Marshall picked flakes of gold from the mill-race. The Mormon workmen followed his example.

When the Mormons did arrive in Salt Lake City, they paid their tithe on their wages, and on the gold they had gathered. It never occurred to them that they had taken Sutter's gold. Nor did Sutter ever collect a penny from the thousands who later mined his property.

The first rich gold ore (highgrade) was found

in Mariposa and in Nevada Counties. In digging out the white quartz ore, the miners often noticed light being reflected from metallic surfaces.

There has never been a miner who could resist the instinctive impulse to pick up and examine the piece of ore which showed the reflection. Ninety-nine times out of every hundred, the metallic reflecting surface consists of a compound of iron and sulfur. This is known to the miners as a pyrite, or sulfide. The non-miner knows it as "fool's gold."

The experienced miner will only drop the worthless material and continue his work. But he still hopes that the next glint will come from real gold.

If that occurs, he has two reasons to rejoice. First, when that hundredth piece of ore is found, it guarantees that the mine will be making a profit from the ore sent to the mill. Thus it will be able to continue to operate, extending his period of employment. But of more importance, if the specks of metal are large enough, or if there are sufficient numbers showing on the surface of the freshly broken rock, it is immediately classified as High Grade ore or simply "Highgrade."

Surely, a very high percentage of the miners working in the rich stopes did pick up pieces of highgrade. They could not resist the temptation to put the richest chunks in their pockets, and begin to look for ways to smuggle them off the mine property.

If the guards did not catch them, they crushed the hard quartz in a steel mortar, panned the resulting mixture of sand and pieces of raw gold, and took the metal to an illegal gold buyer. The buyer paid the miner from eighty to ninety percent of its value. Thus both the miner and the buyer joined the fraternity, known in the gold country, as the Highgraders.

Highgrading was not restricted to stealing rich ore from the stopes. Once the ore reached the surface, many of the mill workers kept a sharp eye focused on the ore as it made its way through the rough crusher, and on into the stamp mill, or into the more sophisticated ball mill.

Once the ore had passed through the complete milling process, the free gold or the amalgam (gold dissolved in mercury) was many times more valuable than the richest ore. It presented an even greater temptation to the men who must have access to it while moving the end product of the mine on its way to the U.S. Mint (the only legal buyer of the newly mined precious metal at that time). If these men took some of the gold for their own use, they too were listed as highgraders.

Many were the methods invented by the surface workers to escape the watchful eyes of the inspectors and the security guards. Consequently, they were watched more closely than were the stopers.

The competition to get mill gold off the mining property was keen. Whereas the richest of the natural ore might be worth up to $100 per pound, the material smuggled from the mill could be valued at $33 per Troy ounce—not $35, as there was always some impurities mixed with the gold—or roughly $400 per pound (12 troy ounces to the pound). If amalgam was taken, its value might be as much as $250 per pound. In either case, it takes only a tiny amount of either material to tip the balance against a one-pound weight.

When, during the depression years, the workers were paid only $5.00 per day, or $125.00 per month, and that new Buick gleamed on the dealer's show room floor, the temptation became very great.

Meanwhile, the mines were not owned or managed by stupid people. Nor were the men they employed as security totally lacking in brains. Hence from the time of the formation of the first mining company, until the gold mines closed when $35.00 would no longer pay the cost of digging gold from the California hills only to bury it once more at Fort Knox, Kentucky, a deadly serious game was played between the mines and an unknown percentage of the mine workers.

This book tells of the various mining methods used, from picking small nuggets from the cracks in the bedrock along the banks of streams with the point of a hunting knife, to the dredgers sorting gold from gravel lying under thousands of acres of rich soil, where the Sierra streams emerged from the mountains.

It describes the mining of the well weathered, "rotted" quartz, honeycombed with threads of gold, being mined by pick and shovel, crushed and panned by hand to recover those threads. It tells of the introduction of mining machinery to help in the recovery of the ore, and to take the gold from it.

Most especially, it tells of the removal of the hardest of rocks from the 500 miles of drifts threading through the Grass Valley hills. Were all these passageways through the rock to be laid end to end, and pointed in a southerly direction, a miner coming to the surface at the end would find himself in a Spanish speaking country; the fast food dispensaries would be featuring tacos and enchiladas.

Endless stories still circulate through the gold country. Only a few could be included in this volume. The story of the mine owner who sold his own gold to the highgrade buyer to escape paying too high an income tax, cannot be told in detail. Too many would still recognize the man who would rather split with the highgrade buyer than with the IRS.

Today, most of the people living in the

foothills of the Sierra look upon the highgraders as but Robin Hoods, taking from the rich to give to the poor, and who was poorer than the hard rock miner's family?

Easy money often leads to greed. Some miners turned to other forms of illegal gains even before the mines closed.

One highgrader quit work in the Empire Mine in 1945. He bought a small mine well back in the hills above Washington. He was able to send a small shipment to the Mint each month. He was satisfied to add a thousand dollars to his income each year by including some illegal gold melted with the gold from his mine. He continued to dig out a little gold even after all the large mines closed. But that is for cover-up only. Now his main business is farming. He tills about one-half acre of bottom land on the creek below his mine. In 1980, his net income from the cultivated plot was $100,000.00!

His cash crop is Marijuana.

FOREBAY, WISE POWERHOUSE. Located on Auburn Creek. One of the first generating plants to use water from old miners ditches.

CHAPTER II

GOLD! THE ROOT OF ALL EVIL?

"Gold is where you find it, and finders is keepers." Or, as the "Cousin Jack" miner firmly believed, "God put hit 'er on hearth for 'em as finds 'it, an' not for the rich bloke what owns the mine."

Historically conquering armies have looted the captured cities, mainly for the women to rape, and for the gold they could carry away. Cortez demanded and got a room full of golden vessels and objects of art for the release of Montezuma, the Aztec ruler and semi-diety whom he was holding for ransom. Drake, Hawkins, and other privateers seeking the capture of this gold on its way to Spain, were the forerunners of a long series of freebooters, who raided the shipping lanes of the world until early in the nineteenth century. They raided, not for slaves and perhaps other cargo of value, as had their piratical ancestors from the Mediterranean, but after the time of Drake at least, they dumped overboard everything, often including the crew of the captured ship, that did not have the immediate value of the gold, the jewels, and if cargo space permitted, the silver.

After Bluebeard and Laffite, the era of the gold grabbing pirates ended. But the glitter and the desire for gold was not lessened. Not much went into history books for some fifty years. Not until the great Gold Rush to California was gold moving freely in commerce again, and thus became fair game for those who would rather risk a shot in the back by a pursuing posse, or a rope around their necks pulled over a stout oak limb, to real labor. Why work when a mask and a trusty Navy .44 pried the strong boxes from the top of a stage coach, or settled any argument as to the real ownership of the "dust" in the Wells Fargo Office?

The period of Black Bart and of Joaquin Murietta did not last long. The miners had worked too hard grubbing that gold from the creeks and from the hillside tunnels to stand for someone getting it "for free." True, they did not hold much for Habeas Corpus, nor for other legal "hocus pocus," and all too often they may have hanged the wrong man, but their method was most effective. The "knights of the road" found it more profitable to deal Black Jack or to run a stud poker game, than to risk getting their neck stretched for using a gun to relieve some miner of the contents of his poke. If all were accounted for, it is almost certain that the highwaymen relieved the coaches and the express offices of no more than a few hundred thousand dollars in dust, sponge gold, and bullion during the first two decades of the gold rush. This was small change indeed compared to the sums taken from the big mines by the highgraders when they really got under way.

The modern bank and armoured car robbers are after currency and negotiable securities, not gold. There was no gold in either the Boston Brinks robbery, nor in the great English train robbery of 1963. Some silver coinage was taken, but it was considered more or less a nuisance. If there had been gold, it would have been buried immediately. It would be much too hard to dispose of to warrant the danger of having it found. It would be a very foolish fence indeed who would attempt to dispose of stolen gold in large amounts today.[1]

---

[1] Since the ownership of gold in the U. S. has been legalized, the ownership of gold is no longer so great a hazard. Now it is but a matter of time before a large gold "heist" takes place.

Gold and the Highgraders

Following the time of the knight of the mask and the Colt, came the time of the highgraders. Briefly, a highgrader is an employee of a mine who appropriates for his own use some of the riches God has put in the rock of the mine in which he works. Presumably he has taken only from the rich mines, a mine under absentee ownership; but all too often he took from any mine, often leading to financial disaster for a struggling mine owner.

Strictly speaking, highgrading refers only to the taking by devious means, of quartz gold ore which has enough free gold showing on its surface to make it worthwhile to be spirited away from the mine, crushed, and the gold it contained recovered by panning. This kind of ore was known as Highgrade.

Eventually the term came to be applied to the illegal removal of gold in any form from the mines. This included everything from robbing the sluice boxes of the great hydraulic mines, to casting a few small buttons of bullion from the mine smeltery and smuggling it off the mine property. It also included the fencing of the stolen gold by anyone. The illegal gold merchant may never have seen a real gold mine, but if he handled stolen gold in any form, he was also a highgrader.

More recently in the mining country of California the term highgrading has come to be applied to any form of stealing. Especially it applies to shoplifting, when the thief limits himself to the stealing of goods of higher quality from the store shelves, while disregarding the more common items.

The universal appeal of gold has given it a special place in the world of crime. Since history began, gold has been demanded by conquerers. Traitors have sold out for nothing less. Almost without exception, invading armies have been preceded by agents carrying bribes and funds for the collaborators. Nearly always this was in the form of golden coins, bars of gold, and golden objects of art convertible into cash immediately.

Alexander bribed the sub-chieftans ahead of his march into Asia Minor and on into India; Hannibal bribed his way across Southern France and on over the Alps; Napoleon paved the roads for his invading armies with golden francs; and our own General Mark Clark was embarrassed by the loss of a hundred bags of twenty-dollar gold coins when the boats from his submarines lying off the coast of Algeria capsized while transferring some ten million dollars ashore to be used to insure the cooperation of the Algerians when the invasion fleet arrived from America and U.S. troops hit the beaches.

It was imperative that gold be used in the African invasion. Even at that late date, the only money those people understood and trusted was good hard gold coinage. Hitler counterfeitted millions in U.S. paper currency and with it undoubtedly bribed many an entry into the more sophisticated world capitals, but the wiley African chieftans would settle for nothing less than the feel of good gold coins between their teeth.

Because gold does not circulate freely in most of the world, the more civilized people can be fooled by bogus coins. Those with experience know by feel and by weight the real from the substitutes. They cannot be tempted by anything but real gold.

One of the great subtleties of the New Testament is missed by most students of religion. Judas was bribed with thirty pieces of silver. Gold was the official bribing medium. Saint Matthew was trying to illustrate how little value man often places on his own integrity. He would, and did, sell his Savior for silver. Not gold, mind you, but lowly silver! The theologian traditionally emphasizes the thirty; Saint Matthew undoubtedly meant that we see that it was only mundane silver that Judas could not refuse!

Gold, because of its retention of color and luster when polished, its slightly greasy feel when cast in large chunks, and its great density, is almost impossible to confuse with any other metal, especially by anyone who has seen it and has handled it in more than minor amounts.

The Empire Mine in Grass Valley was supposed to have had a special brick of gold bullion cast for display in its office during the early part of this century. It was only slightly larger than an ordinary house brick, but as is the case of all large gold bricks, it was cast so that the sides were sloping at an angle of about twenty degrees. In other words the base of the brick was about an inch larger than its top.

The display was placed on a counter top in the office of the mine. Lady visitors were invited to pick it up. They were told that if any one of them could pick it straight up using one hand alone, she could have it! The Mine cheated a little. They made sure that the small side of the brick was always placed up so that a good grip was impossible. Also the mine might well have added a little oil to its surface just to make a good grip impossible. Needless to say no woman ever had the strength in her fingers alone to lift it. Most men could not, as it weighed some sixty pounds. It was worth about $15,000 even at the low price of gold at the time. ($20.67 per Troy ounce)

To demonstrate that the brick was not bolted to the counter top, the guide would slide the brick to the edge of the counter and invite the frustrated women to tip it up. This they could do by getting a hand beneath the end of the brick and tilting it. Most of them would still not believe that it weighed so much. There must have been a trick somewhere.

Perhaps the reader, like the author, will have a little laugh the next time he sees the robber on television or in the movies nonchalantly tossing large bags of gold around!

SETTING MODERN STOPER. BRUNSWICK MINE

CHAPTER III

GOLD IN THE STREAM BEDS

"Gold is where you find it," was the answer the tenderfoot got when he arrived at the diggings and asked: "Where is the gold?" He found the real answer to his question by one of several methods. First, if he had something to live on, he hung around for a few days observing the operations of the more successful "old timers." (They might have arrived as much as three weeks before.) If he was intelligent, he took the best of what he saw and applied those methods to his own use. If he was without funds, and that all too often was the case, he was always able to go to work for one of the successful miners and learn by doing. Then if he could stay away from the card tables, he saved a grub stake to finance his own prospecting. This was, of course, after he had learned what to look for and the most likely places to look.

The stories of gold lying in the streets in California were, to say the least, highly exaggerated. True, nuggets were actually picked out of the gravel used to fill the quagmires which passed for streets in the placer mining towns. Those nuggets were, however, rare exceptions. The early '49ers, as well as those who followed, found that, after all the dangers and the hardships of getting there, they did not automatically become rich. In fact only a very small percentage of them ever did make more than a meager living from the mines.

Almost all great fortunes of early California came from merchandising or from transportation. Only a few of the intelligent men stayed with the mines. For the most part, they soon learned that mining was a dirty, backbreaking, dangerous, and most often unrewarding occupation. They left it for better opportunities.

The Who's Who of early California is filled with the names of men like Stanford, Crocker, Huntington, Hopkins, and Beale; men who began with stage and wagon lines to serve the miners along the mountain streams and who then became merchants and promoters who used their connections to build the Central Pacific Railroad and to establish steamship lines.

On the other hand, men such as General, and later U.S. Senator, John Bidwell of Chico, were on the scene early. They took the easy pickings; in Bidwell's case at least, with the use of Indian laborers from the Chico ranch, then quit while he was ahead. They then retired on what amounted to feudal estates and lived a life which lay midway between that of the great hacienda owners of the Spanish days and that of the English country gentlemen.

The first miners followed Marshall's example at Coloma and worked the river beds only. They found that, although there might be a little gold in the ordinary gravel of the channels along the Yuba, the American, and the Feather Rivers of the north, and in the Merced, the Stanislaus, and the Toulumne Rivers to the south, it was always very fine gold. The real concentrations of the heavy metal were on the bedrock; that is, it lay on the hard, uneroded country rock which formed the walls and the floors of the canyons through which the streams flowed. If the rock happened to be of such consistency that natural cracks and crevices were formed as it weathered, so much the better. The irregularities acted as traps to catch and hold the very heavy particles of gold. The gold, because of its great density, tended to roll along the bottom rather than to be suspended in the rushing water. Consequently, any obstruction on the bottom of the river bed would retard its movement and allow it to accumulate on the upstream side of the rough spots.

As an example, the famous "Tin Cup Mine" at Downeyville, on the Yuba River in Sierra County, produced a fabulous outpouring of the precious metal with a minimum effort on the part of the miners. At that location the Argonauts found a strata of slate rock standing almost vertically, but tipped slightly against the current. Slate formations weather naturally in such a manner that it leaves deep crevices between narrow slabs of hard rock, which, in this case were sticking almost straight up. At the Tin Cup Mine, each of these crevices acted as a catch basin for any heavy metal being carried down the stream. Naturally these cracks were partially filled with sand and gravel, but the gold, being several times heavier than the ordinary materials being moved by the running water, insidiously worked its way to the deepest part of the depressions as the current constantly shifted its course and changed its velocity.

The mine reportedly got its name from the habit of the miners working that stretch of the river to wade out into the icy water, scoop up a bucket of gravel from between the slate projections, wade back to shore, and pan to recover the gold. The gravel was so rich that it took only a few hours to fill their tin coffee cups with dust and nuggets. Since this was worth about $2,500, even at the low price paid by the local gold buyer, each miner figured that he had had enough of the icy water for one day and promptly adjourned to the nearest saloon, where many and varied were the devices presented to help him get rid of the fortune so easily acquired. If he could resist the temptation of the "Rot-Gut" whiskey, Faro, Black Jack, and Stud Poker; if he could keep out of the clutches of the gamblers (not necessarily crooked, but it was quicker that way), there were always plenty of girls of assorted sizes, colors, and shapes to help him forget the cold water, the poor food, and most of all, the terrible loneliness of his separation from family and friends.

Most of the so-called female entertainers relieved the "poor" miner of his dust by "doing what came naturally," but they were also well trained in the art of picking pockets and wheedling the customer out of an extra fee for her "poor sick mother," for whom she had entered this sordid life. But if nothing else parted the sucker from his wealth fast enough, she was not above slugging him, or putting a few drops of laudanum in his drink to render him senseless until she could remove everything of value from his person and roll him out the door. He woke up, if he was lucky, a sadder and, we hope, a wiser man.[1]

Not too many places with natural riffles to trap the gold, such as the slate layers at the Tin Cup Mine, were found. Mostly, the gold was on the bedrock under the ordinary gravel bars. Bars were the long tongues of sand, gravel, and boulders deposited as the streams changed their courses or changed their grades, slowing the current causing them to drop their heavy loads. As the water decreased in velocity the larger and heavier chunks of gold (nuggets) were dropped first. As the stream continued to run more slowly, smaller bits of gold were put down, mixed with the more dense parts of the stream-worn gravel. Finally, the very fine flakes of gold were deposited along

---

[1]To the surprise of the modern reader, the girls almost without exception married, raised a family and settled down to the life of a housewife.

with the ordinary gravel. There it remained in a more or less uniform mixture until carried along by the next spring flood, or until the restless erosion of the stream cut the bedrock back from the next falls or rapids below. Then the sorting process began all over again.

Often the stream changed its course so much and so fast that a tongue of gravel would be isolated and no longer subject to the erosion of the stream. As the water cut deeper along its new course, the isolated bar was left higher and higher on the canyon wall. Usually the lateral erosion of these canyon walls allowed the gravel to wash down the canyon walls to eventually join the main stream once more. If, however, the bar was left intact, or at least partially intact, it now became a bench. Due to its distance from the stream below, the bench was usually hard to mine. The first miners simply carried the bench gravel down to the water in the stream below, where they washed it for its gold content.

The carrying process was both backbreaking and slow. The average miner certainly could not have been called lazy. The hard work did not bother him, but he was in a hurry; so by the summer of 1850, the miners decided that, "if the gravel would not go to the water, they would bring the water to the gravel."

So began the series of ditches and flumes which interlaced the lower mountains and foothills of central and northern California, and which later became the nucleus of many of our irrigation systems, as well as the basis for the water power for electric generating stations.

Some of the earliest power houses are still in operation. The P.G. & E. is still generating power from the force of the falling water carried in the old miners' ditches, then selling the water, now stripped of its kenetic energy, to the ranches below. Until the recent change in the alignment of Interstate 80 just below the city of Auburn, the motorist drove within a few hundred feet of such a generating plant (The Wise Power House). He could see the water, now divested of its energy, flowing quietly into an irrigation ditch to be carried to, and distributed in, the deciduous fruit orchards around Newcastle, Penryn, Rocklin, and on down to Roseville.

The water to operate the Wise Power House may be seen in the large, swiftly flowing ditch which crosses under State Route 49 just north of Auburn. This ditch follows the natural contour of the land to the edge of Auburn Ravine. There it is trapped and forced into penstocks which plunge down the steep slopes where the water under pressure pushes against the blades of the turbines. These turn the generators (alternators) which convert the energy into electrical current. The ditch was dug originally to bring water to the gold rich gravel on the bottom and along the sides of Auburn Ravine. It was dry except during the wettest of the winter months.

EMPIRE MINE. Catwalk, Office to Mill

CHAPTER IV

THE FIRST ARRIVALS

The newly arrived miner soon learned the basic facts of placer mining. Then, having provided himself with a pick, a shovel, and a gold pan, he set out to make his fortune. Usually he just moved up or down the stream he watched being worked until he arrived at a spot not taken up by some earlier arrival. Here he staked his claim. This he did by driving a stake, with his initials carved on both sides, at one edge of the ground he wanted to work. He then measured off the number of feet allowed in that particular mining district along the stream channel, then drove another initialed stake. He recorded his claim with the local mining committee, or with the town clerk, or still later with the County Clerk when such offices were established. Such a procedure is still in effect today, provided, of course, the land is in the Public Domain and open to mineral claim entry.

If the pressure from other miners was not too great, the new miner would spend some time sampling his claim before staking it to see if it contained enough gold to warrant claiming and recording. If he got nothing, or only a few colors, he moved on to the next likely place. The inexperienced miner often missed rich deposits because he was not adept at the art of panning, or because he failed to take proper samples. In some rich ravines almost any sample of the dirt or gravel showed some flakes of gold, or at least the crease of the pan showed a considerable deposit of "black sand" (magnetic iron). This would indicate the presence of mineral bearing gravel, and the probable presence of gold. In other locations it was necessary to take samples from close to the bedrock, or from immediately back of an outcropping where the gold was trapped as it was being moved down stream.

Such was the case at Auburn Ravine. The first two parties to prospect the area each found gold but in insufficient amounts to stop them from going on down into the canyon of the North Fork of the American River. There they knew they could make from $10.00 to $20.00 per day ground sluicing. The third party dug at the right spot and panned properly. Within three years almost a hundred thousand ounces of coarse gold had been washed out. Single pans of from fifty to one hundred dollars were not uncommon. The gold had been there, but it took skill and know-how to find it.

Paraphrasing W.B. Lardner, in his History of Placer County, 1924, we find that the Auburn Ravine was first known as "North Fork Dry Diggings." In May, 1848, Claude Chana, a French Canadian, in cutting across the foothills from Johnson's Ranch on the Bear River, to Coloma on the American, camped here and found gold, but moved on to what he thought would be richer diggings. The Ravine was next prospected in August, 1848, but the names of the miners have been lost. Apparently they did not wish to be remembered for having overlooked the great bonanza.

The area next became known as "Soldiers Springs" because it was mined in the spring of 1849, by discharged soldiers from Colonel Stephenson's Regiment.[1] The soldiers mined for a

---

[1]The regiment had been formed in New York and shipped around the Horn to California to help the Americans there fight in the Mexican War. Arriving too late, they were discharged and told to make their way home as best they could.

short time, then were stampeded to the Yuba River by the rumors that fabulous nuggets could be picked up there right off the surface of the sand bars. That is, they all left except John S. Wood. He was tired of chasing the rainbow. He settled down, built a cabin and continued to mine. The Ravine now became known as "Wood's Dry Diggings." Suddenly some miner showed up with a long tom. When it was put into operation, fortunes began to roll in. How the name Auburn came to be attached to the town which developed is not known, but John A. Markle, in his diary, speaks of spending the night of November 15, 1849, in a tent in "Auburn."

Such is the history of a typical mining camp which turned into one of the charming little cities of the California foothills.

Most of the miners arrived broke, they had no pick, no shovel, no gold pan, and in fact, no idea of how to use one if it became available. At first there were no jobs even with successful miners. The feeling was, however, that there was gold enough for all; so the new arrival was shown how to dig dirt and gravel from the cracks and crevices in the bedrock where it protruded above the water, or from the dirt along the stream banks.

The new miner, with nothing but his hunting knife and his hat or an old sock to be used as a container, scrambled along the rock ledges above the water and cleaned out the crevices with the point of his knife. When his container was full, he carried it down to the water; where, with the tin plate from his pack, he proceeded to separate the gold from the base materials. He soon found the trick of mixing just the right amount of water, and of giving the pan exactly the right circular motion as he agitated the contents to give the gold a chance to settle to the bottom. Later he found that if he very carefully cleaned out his frying pan to remove all the grease, it could be used as a gold pan. With greater capacity and with its sloping sides, it was much more effective than the plate.

Naturally this was slow, laborous, and not too remunerative work. But with any luck at all, the newcomer could make from three to five dollars per day by this primitive method of "sniping." Since living cost almost nothing, he was able to acquire the proper tools within a relatively short time. True, flour sold for a dollar a pound, beans and bacon were correspondingly high, but the new miner literally lived off the country. He had his rifle. Game was still plentiful, and the streams where they were not fouled by the miners, were still full of migrating salmon and other fish. They could be picked out of the riffles with the bare hands! Any ambitious man could live by this means alone until he could dig out enough gold to purchase the tools of his trade and devote all of his time to mining.

The very first Californians to arrive on the American River near Coloma followed Marshall's example. They merely diverted the water from the river, and used the force of the running water to sweep away the dirt, sand, and gravel which covered the bedrock alongside the main channel. After bedrock was exposed, the water was turned off and the miners proceeded to pick up the nuggets which had not been washed away. They soon found that most of the gold could not be picked up directly, but must be scraped out of the cracks in the bedrock, and then separated by panning. This method came to be known as ground sluicing. With some modifications it is still in use today in primitive areas.

Not one of the first Californians arriving at Coloma had ever had any mining experience, nor had they seen a mine of any kind in operation. They were, however, practical men, and while they had never heard of the science of physics nor of hydraulics, they were quick to figure that if the heavy gold in the stream beds was trapped by the irregularities on the bottom, it would be wise to make more irregularities. So they threw obstructions across the current below the area being sluiced. They found that some of the gold which had escaped them before was now trapped. With this experience as a background, some smart miner realized that if the gold bearing gravel was moved through a long box with obstructions built across the bottom, the heavy materials would be trapped more successfully.

Long, open-top boxes were built. One-inch cleats were nailed across the bottom at ten-inch intervals. Gravel was shoveled into the box at its upper end, a stream of water was fed into the box, and the washed gravel (the tailings) were directed back into the stream below. The width and depth of the box depended on the amount of water available, the steepness of the grade at which the box could be set, and the size of the largest stones in the gravel to be handled.

Gold and other heavy materials were stopped on the upper sides of the cross cleats (riffles). The clean-up was accomplished by turning off the flow of water and scraping the materials from the upper side of the cleats with a square point shovel. The separation of the gold was still a problem. It was solved by one of the first sailors to jump ship in San Francisco Bay and rush to the new mines.

It is not generally known that some gold was found in the streams of the Great Smoky Mountains in the Carolinas in early colonial days. It was mined intermittently for some two hundred years by the Scotch-English ancestors of Daniel Boone and his kinfolk.

Not all the surplus population of the mountain hollows followed Boone to the valleys of Kentucky and Ohio. In the mid-1840's, some of the more independent broke with tradition and moved east. In Charlston, they answered the "call of the sea" and eventually found themselves on San Francisco Bay in the spring of 1848. They, and every other sailor, fell prey to the call of the golden siren. They deserted their ships at the first opportunity and rushed to the hills. They were not alone. With no men to work the ships, the captains and all the other officers saw no reason to remain behind and allow the common sailors to get rich while they acted only as watchmen. They too took to the ship's boats, rowed up the Sacramento and the San Joaquin Rivers to Sacramento and to French Camp, and only a short time later, to Stockton and to Marysville when settlements developed at those strategic locations.

The ship's officers were often joined by the army officers from the Presidio in San Francisco. The common soldiers had long since deserted and rushed to the hills along with most of the male population of the sea port towns.

One of the "hillbilly" boys to desert his ship in favor of the life of a gold miner, had once helped his "pappy" wash gravel for placer gold during the winter when there was no corn or tobacco to hoe. With him came the gold pan. The Tarheel, ex-farmer, ex-miner, ex-sailor, and now modern Argonaut, realized at once that he knew something that would give him a head start on the other new arrivals. History does not record where he got the thin sheet-iron from which to form the first pan, probably from Sutter's Fort on the banks of the Sacramento River; or if he had the help of a blacksmith in hammering out the first pan with its sloping sides; nor do we know whether or not he sold any of his pans to his fellow miners. The chances are that he was so infected by the gold fever that he made one pan for himself. Others watched his use of this improved method of separating the gold from the fine sands. They realized that he had solved one of their main problems. They got a blacksmith, who was not doing too well at mining, to hammer out one for them.

We can only hope that the sailor-turned-miner found enough gold to make it worthwhile. He certainly passed up an opportunity to make a fortune by cornering all the sheet iron in California and selling pans and instructing the greenhorns in their use.

At about the time the Carolina sailor introduced the gold pan into the camps along the American, the Spanish-American miners from the Newhall area, just north and west of the village of Los Angeles, heard of the rich discoveries in northern California, and set out to investigate. They did not push on to the proven mines on the American, but stopped to prospect the Merced and the Toulumne Rivers. On both rivers they found good diggings and settled down to exploit their good fortune before the Gringos arrived to start competing for the gold in the benches and on the bars.

The Mexicans had mined in southern California and in northern Mexico; so this was "old hat" to them. The only difference was that here they could recover as much gold in a day as they could in a whole month of hard labor in the mines to the south.

The Mexicans' first contribution to the mining industry was the separation bowl (the batea). This was a wooden bowl about twelve inches across and five inches deep. It was carved from a single piece of myrtle or oak wood. The method of operation was essentially the same as used with the iron pan. The same swirling motion was used to bring the lighter materials to the top and to hurry the heavier particles on their way to the bottom. Because the sides were steeper, the bowl was held at a sharper angle while being used. The Spanish Californians also contributed the "arrastre" to the mining industry, but that came later, after the discovery of gold in quartz.

At this point it is interesting to note that John Bidwell missed the honor of being the discoverer of gold in California because he did not have the funds needed to send to Old Mexico to purchase a batea! In 1844, a Mexican ranch hand working for Sutter under Bidwell's supervision, told him that there was much gold in the black sands of the Bear River. During a trip from Sutter's Fort to the "Hock Farm" on the Feather River, they had paused for lunch on the banks of the Bear. The Mexican pointed to the black sand back of a rock in a riffle and told Bidwell that the gold was there, but that it could only be recovered by the use of the batea, which only the experienced miners in Mexico could make. He alone knew where to go to get one.

It was agreed that they would say nothing about the discovery until Bidwell could raise the funds needed to send his companion south for the necessary mining equipment. Meanwhile an Indian rebellion broke out. Bidwell, who was already commissioned a Major in the Mexican Army, was called up to help put down the Indians. He was thus occupied off-and-on for three years. He thought no more of the gold until after Marshall made his world-shaking discovery at Coloma, and demonstrated that the Mexican batea was not necessary. In June, 1848, Bidwell rode back to the crossing of the Bear where he had seen the black sand. He was too late. A hundred red-shirted miners were already taking a hundred dollars a day from the river sands!

CHAPTER V

PLACER MINING

The ability of gold to work its way deep into the cracks of seemingly solid rock is illustrated by the following story.

In April of 1849, there arrived in the new Grass Valley diggings, one Michael McGuire. He had come from Ireland to Pennsylvania where he started to farm. When the news of the California gold strike reached Philadelphia, he kissed his wife of only a year goodbye and sailed on the first ship to clear for Panama. He walked across the Isthmus, through the disease-riddled jungle to the Pacific Coast, and used some of his Irish "blarney" to talk his way aboard the first ship that touched the port of Panama City. He arrived in San Francisco without having contracted either malaria or yellow fever. In the Bay, he found a small boat going up river as far as the new town of Marysville. From there he shouldered his pack and walked to the gold fields on Wolf Creek. He staked a claim on a branch of that creek, one mile east of Boston Ravine and Gold Hill, where gold in quartz was found only eighteen months later.

He worked his claim until the water stopped flowing during the long dry summer. He accumulated several thousand dollars in coarse gold, so he registered his claim with the local mining committee, and retraced his steps to Philadelphia. There he sold the farm, and with his wife and new born daughter, he set out for California, secure in the knowledge that if the gold did run out, he must settle in the land of sunshine and opportunity.

They arrived back at the claim in time to

mine during a part of the new rainy season. Michael again washed out enough gold to carry his family during the succeeding dry season. He used the summer to build a cabin and begin the clearing of what was to become a thirty-acre ranch and orchard. During the next twenty years his routine was to mine by ground sluicing during the winter and late spring when water was available, and to raise fruit, vegetables, and children during the remainder of the year. In all, his Irish Coleen bore him ten healthy children; seven boys and three girls, and wonder of wonders, in spite of childhood diseases, hunting accidents, rattle snakes, caches of mine explosives left fully exposed in the immediate area, and open mine shafts abandoned in the woods where the children played, they all survived to maturity!

Almost at once McGuire had "a better idea." By prospecting his claim, he found that he had about $100,000 to be mined. Unlike all the others, he did not dig it all immediately and lose it in an unwise investment or in the failure of the banks. His wealth was in the safest bank in the world. His gravel bank! He washed only enough gravel each year to live comfortably. Only one concession was made. As each of the children was born, he worked a few more days and took out another thousand dollars which he put in a savings account to be used for the education of that child.

Sad to report, when Michael, Jr., the youngest child, was born, and the usual $1,000 deposited, the bank defaulted during the panic of 1873 and the entire amount was lost. They did not panic. There was still enough gold left to send those who wished to go to the new University in Berkeley.

The fifth son, James, graduated from the University with degrees in civil and mining engineering. He then migrated to South Africa, where due to his experience in the Grass Valley mines [the most progressive in the world at that time] and to his scientific knowledge, he immediately became one of the leaders in the development of the Rand Reef Mining District.

In April, 1912, James was invited to speak at the World Mining Congress in London. As he was in the northern hemisphere, he decided to cross the Atlantic and visit his aged mother. After the speech, he boarded the boat train for Southhampton. The train was delayed by a small derailment which caused him to miss his ship. The White Star Line, however, put him aboard another ship which was sailing almost immediately. He arrived in New York in time to keep his reservations on the pullman train across the country. He stepped off the Narrow Gauge passenger car in Grass Valley to find that there was no one there to meet him. He thought this most unusual, but he walked the five blocks up Main Street to his mother's house. There, without knocking, he walked in, only to find that the entire family was gathered in the parlor to decide what form of memorial service thay should hold for him. He was supposed to be dead! The ship he had missed was the Titanic!

James returned to Africa, but when it became apparent that the United States would become involved in World War I, he was called to Washington by President Wilson. There he became a "Dollar a Year Man," and served as a presidential advisor on strategic metals. After the war, he moved to New York where he managed the mining portfolio of Bernard Baruch, the "advisor of presidents."

The sixth son, Thomas, also graduated from the School of Mines in Berkeley. He followed the profession of civil and mining engineering in Grass Valley until his death in 1968, at the age of 93. At that time he was the last living man who had ever held the "swing bar" of a great hydraulic monitor as it blasted mountains of gravel off the hillsides and started them on their journey to the Sacramento Valley, there to cover whole towns and living orchards.

The point of the story is that Michael McGuire, Sr., carefully worked and reworked the gravel along the creek which crossed his claim on Little Wolf Creek, right down to the bedrock. This he carefully scraped to get every speck of gold out of every crack and crevice. The original claim along with the surrounding land was located under the Homestead Act of 1862, and was patented as agricultural land. The youngest son, Mike, worked the old homestead as a dairy ranch during World War I and on to the year 1924. The ranch was then abandoned as farm land and was allowed to revert to forestry. The old orchard which had supplied the family for three-quarters of a century, has withstood the encroachment of the pines and the firs, and still [in 1979] produces some apples and pears from trees which have had no care nor irrigation for more than fifty years.

In 1936, during the Depression, Mike, Jr., was managing the old ranch for the family. There was still some income from the rental of the house, from the sale of cordwood, and from the sale of stumpage for saw timber. He was approached by a sniper, (a miner who was searching for overlooked gold deposits) who wanted to rework the creek bed on the original mining claim. Mike thought he was crazy. He knew how thoroughly his father had worked the ground, and he thought of the many hours he and his brothers had put in scraping the bedrock once again to pick up a dollar to spend in town on a Saturday night. If there was any gold there, Mike felt that he should know about it if anyone did. He was a pretty good miner himself. He had worked most of his life in the mines in every capacity from "mucker" (rock shoveler) to superintendent. But he told the sniper to go ahead, but warned him that there was nothing there. The usual royalty of ten percent was agreed on.

Mike and every other mining man in town was struck speechless when, at the end of the first month the sniper came in, counted out ten crisp new ten-dollar bills and asked for a receipt for the royalty on $1,000 in gold he had sent to the Mint! This was repeated on three successive months.

Needless to say, a mild gold rush started. Every sniper in town except those that were sure that this was just another scheme for getting rid of highgraded gold, was out the next morning to see what was going on.

The explanation was simple. During the eighty years between McGuire, Senior's, activity on that part of the creek, and 1936, Mother Nature took over and softened the bedrock by chemical weathering. By using a sharp pick made from much harder steel than had been available before 1900, the sniper was able to chip away another half inch of the dioritic-porphoritic bedrock. He carefully swept up the chipped material and panned it. The extremely fine gold (flour gold) had worked down into the cracks which would not have been visible had not the iron which followed the gold, stained the surrounding rock. These cracks were so loaded with the flour gold that each pan gave up from three to five dollars.

The bare rock where this took place can still be seen today. It is about one thousand yards south of the Union Hill School on the Grass Valley-Colfax road (State Route 174). It lies at the easterly curve of the road just at the intersection with the dirt road which runs to the top of Osborn Hill. It is also only a little more than a half mile southeast of the Empire Mine.

Another example of the penetrating power of

gold due to its great density was demonstrated by a mining venture on the Klamath River in Siskiyou County, close to the California-Oregon border. There, due to an oddity of nature, a layer of relatively soft clay was sandwiched between two strata of slate which, due to the movement of the earth's crust, had been tipped to a vertical position and eroded away by the rushing waters of the river. The clay, being so much softer than the slate, was always eroded to a depth of several inches below the slate on either side. This formed an effective trap for the gold which rolled along the stream bottom. Of course the slate wore down, and one would think that the gold would once again be carried on down stream. But such was not the case. The gold was so much heavier than the clay, that even though the clay was very compact, the gold sank deeper and deeper between the slate layers. The first miners on the river found this deposit and dug out several thousand dollars worth of pigeon-egg sized nuggets. Then they were forced to give up mining on the strange ledge because they could not recover the gold once they had mined out the clay. In their pans and their sluice boxes the clay just "balled-up" and instead of releasing the gold it contained, it managed to pick up and hold tight the gold which had been broken loose by mechanical means.

During the Depression of the '30's, however, someone discovered that the clay, if mixed with water and boiled in a large "soap kettle," slaked because of its lime contents. It now became a soupy mixture from which the nuggets and even the finest of the gold was easily panned. The clay layer was worked to a depth of more than 25 feet before no more gold was found. The deeper the miners worked, the more of the gold was found in the form of nuggets up to the weight of thirty ounces. Almost no fine gold was found below the fifteen-foot mark. There is only one way that gold could have gotten in that clay layer in the condition in which it was found. The great difference in the density of the gold and of the clay had allowed the heavy metal to work downward through the clay and thus escape the eroding power of the stream. Its rate of sinking was always just a little faster than the rate of erosion of the rock on either side.

That most of the gold was to be found at, or very close to the bedrock, was, of course, learned by the first of the placer miners after only a few days of work. The fact was farther illustrated by the "Scott Bar Incident." Scott Bar (the bar came from the gravel along the Scott River, and not from a drinking place), was a booming town on the Scott River about a mile above its junction with the Klamath, in Siskiyou County. At that point the river made a sharp bend and in doing so deposited on the inside of the curve a pile of gravel which covered some two hundred acres.

The outside of the current was undercutting a formation known as "Quartz Hill." A large white quartz ledge was being eroded away, and could be seen about two hundred feet on the hillside above the river. The quartz ledge carried occasional chunks of iron sulfide, the cubes of which were sometimes laminated with plates of pure gold.[1]

During the ages, this ledge had been eroded away by the elements. Most of the high grade sulfides remained on the bottom of the river where they were covered with gravel, as the river worked

---

[1] In 1935, George Milne, of Fort Jones, worked the ledge. He uncovered one piece of quartz, almost exactly one cubic foot in size. It consisted of about one-half plain quartz. The other half was made up of one-inch cubes of pyrite inlaid with gold. The rock should have been saved for mineral specimens, but unfortunately it was broken up for the $10,000 of gold it contained.

its way eastward through Quartz Hill. In doing so it mined more of the quartz ledge. The iron in the iron sulfide gradually oxidized. This released the gold layers which, being heavy, naturally remained on the bedrock.

The rich deposit was discovered in 1850. A wild rush soon populated the new town with 3,000 men and 30 prostitutes. Some time later families began to arrive. Homes were built along with a post office and numerous stores. A Masonic Lodge was issued a charter, the second in California; and a Masonic Hall and several churches were soon under construction. The homes and the stores were built on worked-over land, but the public buildings rested on foundations which stood on the gravel of virgin ground.

This situation existed until 1933, when Herbert Hoover and some of his associates spent about $150,000 to build a five-mile flume to bring water to a point above the town site. Their announced intention was to ground sluice the supposedly virgin gravel from under the crumbling buildings that had occupied the area for more than 80 years. After completing the flume and building the sluice boxes to catch the gold, the Hoover group washed gravel for one day only! Then they pulled up stakes and moved out. They sold their bulldozer, trucks, etc., for ten cents on the dollar, and made no more effort to mine.[2]

Gradually the truth leaked out. It seemed that the Hoover people had been "out smarted" by the supposedly "stupid" Chinese. The Oriental miners always followed the Americans to clean up what they had missed in their hurry to grab what they could, then hurry on to the next strike. The

---

[2]George Milne bought the Hoover machinery and used it to prospect the Gold Hill ledge. He paid for the machines from the one piece of quartz alone.

Chinese had tunneled down through the kitchen floors of their cabins located on the edge of town. Then they followed the bedrock to the unmined areas. They timbered very carefully so that the caving in of the gravel above would not give them away, and proceeded to carry out the gravel which contained the gold. This they panned by the river where they were reworking the tailings from the earlier mines. No one was the wiser.

At first the Siskiyou natives had a good belly laugh at the thought of the ignorant Chinese hoodwinking the world renowned mining engineer. However there soon began to be serious doubt as to just who had outsmarted whom?

It seemed that by the time the Hoover Group was ready to turn on the water and commence their mining, the Governor of the State of California signed a bill which made it illegal for placer miners to block off the streams where they were working. They must allow the fishermen free access to the water. It just so happened that this particular stretch of the Scott River was one of the best steelhead fishing spots on the whole of the Pacific Coast! Until the Governor signed that bill, it was possible for the mining company to fence off the stream and to keep the public out as long as it mined or pretended to mine.

It is an historical fact that Hoover's favorite form of recreation was steelhead fishing. With the possibility of reserving the stream for his exclusive use removed, he stopped his pretext of mining and pulled out.[3]

The local people now believe that the mining

---

[3]The writer has fished that stretch of the Scott River many times. He can understand why Hoover was so anxious to get control of those riffles. The Governor who took that privilege away from him was a Republican!

was a "blind" to get control of the river. After all, it does not seem possible that one of the world's greatest mining engineers would risk his reputation and his own money as well, when a few holes drilled by a dredger prospecting crew would have produced the samples which would have shown the gold gone. True, the samples would not have produced the artifacts which proved that the Chinese had been the ingenious culprits, but $150,000 would seem to have been a rather high price for a few rotted timbers, and candle holders, picks and shovels, of the type used by the Oriental miners of the 1860's.

Sluicing Near Nevada City, 1852

— Photo Courtesy of California State Library

CHAPTER VI

THE FIRST MINING MACHINES

Panning was a slow backbreaking job. It could be used only where the gravel was very rich, or following the collection of the sands after ground sluicing. Of course only a small fraction of the gravel in the canyons lay in positions where it could be washed by large volumes of running water; so it was obvious that other means had to be found, or mining must stop. In addition, long periods of panning was conducive to rheumatism. The panners' hands were constantly in cold water, and as rubber boots had not as yet been invented, his feet were wet most of the time. Ten men died from rheumatic fever and from pneumonia and related diseases for each one who died a violent death in the mines.

The Mexican miners used a wooden box about one foot square and five inches deep. Its bottom consisted of a screen of some type which allowed only sand, small stones, and hopefully, the gold to go through. The box was loaded with gravel, picked up and shaken. After the small material had passed through the screen, the rejected material was thrown away. The fine materials were panned.

Isaac Humphrey, a Yankee ship's carpenter coming off a beached ship in San Fraancisco Bay, built the first gold rocker or cradle. To get away from the strenuous work of lifting and shaking the screened box, he cut out a pair of curved boards shaped like the rocking boards under a child's cradle. He attached the rockers to the Mexican's box, then found a large flat rock for a base and rocked the box back and forth to produce the equivalent of the shaking motion the Mexicans used. Of course it did not take long for someone to figure out that a little water poured in with the gravel would speed up the screening. So the miner shoveled the rocker full of gravel, moved it back and forth with one hand and dipped up water from the stream or pond and poured it into the box. As the water came through the screen it carried the accompanying sand (and gold) away from the base of the cradle to a settling basin provided to catch the screened materials.

Someone noticed that the heavy gold settled closest to the cradle. Any irregularities on the floor of the catch basin speeded the process. This naturally led to the idea of setting the rockers on a wooden platform which sloped gently towards the catch basin. Half-inch cleats running across the slope were nailed to the platforms, and the screened materials forced to run over them on their way to the basin below.

This worked rather well at first. Then someone reasoned that if the whole platform moved from side to side the settling process of the gold would be more efficient. So the rockers were removed from below the screening box and placed under the cleated platform. The screening box was then fixed about 16 inches above the platform. Now the entire assembly rocked back and forth. It was also found that if the device was operated with short jerky motions rather than by long smooth swings, better separation of the gold from the sand could be accomplished.

This completed the development of the rocker, or if one prefers, the cradle. It is used today with only minor variations. It is so simple that it can be constructed on the spot from primitive materials. If a metal screen is not available, a screen can be made by boring holes of appropriate sizes in a board nailed to the bottom of the box. If a suitable board is not at hand, a screen can be made by peeling willow shoots and weaving them into a mat with one-quarter to one-half inch openings between the meshes. As a substitute for the metal screen it will not last long, but it is easily replaced and the materials are almost always close at hand on the banks of the stream where the prospector is working. The boards for building the box, the platform, and the rockers may be made, as did the pioneers, by hewing them from a convenient tree with a hand axe. A few nails are usually needed, but any really handy man can fit the whole thing together with wooden pegs.

As the readily available supply of rich gravel began to run out, the miners looked for ways to recover the gold from the larger deposits of gravel in the river canyons. To handle the larger volume which must be worked, the "Long Tom" was invented. This was in reality just a small sluice box. The part which made it different was that the upper ten feet or so were "double decked." This upper deck was fitted with a screen to form its bottom over the entire length. The lower "box" was fitted with the usual cleats to make the obstructions to retard the movement of the gold down its length that extended well beyond the end of the upper box.

The Long Tom did not move. To operate it, one man shoveled gravel into the upper end of the upper box. A second man dipped up the necessary water, or regulated its flow if a stream was available and had been led in to mix with and move the gravel along. He also stirred the gravel with an ordinary stable fork and used it to throw out the larger stones. At the lower end of the upper half of the Long Tom, the gravel which had not gone through the screen, or had not been thrown out by the stable fork, was thrown out with a flat point shovel.

At the end of each day, the "concentrates" from the back side of each cleat in the lower box were collected and panned.

A French-Canadian by the name of Baptiste has been credited with the perfection of the Long Tom. He had left the fur trappers of the Hudson Bay Company after an unsuccessful season of beaver trapping in northern California to go on south to the state of Sonora to placer mine. He left his Mexican wife in the spring of 1849 to join his neighbors who had heard of the rich diggings to the north. He saw the need of something larger than the rocker, so he capitalized on the double deck idea and built the more efficient device. Just where the name came from is not known. Most certainly it did not come from its inventor. There is just no way that his first name could have been Tom!

The main advantage of the Long Tom was its efficiency. With a pan alone, one man could process only a few cubic feet of gravel in a ten-hour working day. With a rocker, by working really hard, he could process and pan one whole yard per day, but with a Long Tom, two men working together could work five or more yards. The actual amount would depend on the gravel not being compacted, and being of small enough average size to run through the box easily. The availability of a stream of clear water was also a controlling factor.

If there was a sufficient head of water to operate the ordinary sluice box it was almost always used. It was so much more simple to operate. In fact one man working alone could put through several yards of gravel in one day. It all depended on the amount of water available and, too, just how much gravel the miner could move into the box during the day's work.

There was one drawback to the sluice box even where the water was available and the gravel could be shoveled in with ease. To recover all, or nearly all of the gold, the sluice boxes were often made hundreds of feet long. In this case, it became impractical to clean-up at the end of each day's work. Quite often, when water was in sufficient volume to operate a sluice box efficiently, the recovery of the gold was postponed until the end of the week, or even until the end of the month.

With long unguarded sluice boxes open to the public at night, robbing sluice boxes soon became a favorite nocturnal occupation of "gentlemen" who objected to the strenuous work of operating their own claims, and who were not skillful enough to make a living at the Faro or the stud poker table.

CORING MACHINES NOW DRILL VERTICAL SHAFTS. CORE COMES OUT IN LARGE ROUND BLOCKS. THIS IS GREEN SERPENTINF ROCK FROM BRUNSWICK MINF

CHAPTER VII

GOLD RUSHES

The Rush of the '49ers was so spectacular that all other gold rushes the world over are dwarfed by comparison. No other voluntary, unorganized migration in history has ever come close to that to California between 1849 and 1855. In 1849 alone, well over 100,000 arrived by boat via Panama, or around the Horn, or by wagon train across the plains and over the mountains. All the newcomers to the gold fields are referred to as the '49ers as not more than 5,000 came during the year of 1848. These came from the Sandwich Islands (Hawaiian Islands) and down from Oregon, or up from Mexico. Before the end of that first year, there were also 1,000 or more from the west coast of South America, and even a few from the teeming cities of the coast of China. The main migration of the Chinese began at the end of the Civil War, brought on by the demand for cheap labor for the construction of the Central Pacific Railroad; but even before that, some 10,000 had crossed the Pacific in search of a better life as miners or to serve the wants of the miners.

By the year 1855, the population of California had reached 500,000. After that, the flow of disappointed miners back to "civilization" tended to balance the inflow. Thereafter, California enjoyed only a small but steady growth until the "Okies" streamed in to escape the Dust Bowl of the depression years. That great migration was continued by the rush to the shipyards and to the aircraft factories during the war; a rush which has not as yet diminished to any extent. It averaged about 350,000 per year between 1941 and 1964, and a quarter of a million a year since that date.

Minor rushes developed within California itself. First there was the ill-fated Gold Lake rush of 1850. Fortunately, no more than 100 miners followed the trapper who had wandered down out of the high country on the headwaters of the Yuba River to show a bag of nuggets he claimed he had picked off the bottom of a beautiful little mountain lake. He drank up most of the nuggets, then set out on a return trip to get more. Other miners followed to find the source of his wealth. He was never able to find the fabulous lake again.

In 1855, some one with a trading post on the Kern River east of Bakersfield sent a report to the San Francisco, Stockton, and Sacramento papers, telling of rich strikes of coarse gold in the gravel found near the headwaters of that rugged stream. Supposedly $10.00 to $100.00 pans were to be found in the creeks flowing down the south-west slopes of Mt. Whitney. Indeed some gold was found, but 10 cent pans would have been a high average in the limited amount of gravel available.

Nevertheless, more than 5,000 skilled miners dropped their tools in the proven mining districts, converted all their assets into cash and took off after the phantom gold of the southern Sierra. This was a real blow to the whole economy of the mining area. These were not just 5,000 men out of a population of 250,000 in the Sierra foothills. They were, for the most part, the key personnel from the producing mines, or they were the owners of claims producing from $10.00 to $25.00 per man per day in new wealth.

Soon the wholesale houses in Sacramento and in San Francisco felt the impact of the reduced purchasing power of the mining towns; then the foundrys and the machine shops experienced reduced orders, and finally the transportation companies were hard hit.

So one greedy store keeper on the Kern was able to start a chain of events which led to a major recession; and an even worse depression was averted only by the good sense of the miners who saw that they had been hoodwinked and pulled out immediately. They returned to their former jobs and to their abandoned diggings. The golden flow of dust and amalgam began once again.

Quickly following the Kern debacle came the equally disasterous rush to the Fraser River in British Columbia. This was not only costly from an economic point of view, but was costly in human life as well.

As many as could find shipboard space sailed from San Francisco to Vancouver. There they took small boats and headed up the river to the mines. Most of the 10,000 or more who took off in search of adventure and easy wealth chose this way of getting to the diggings. Several overloaded and unseaworthy ships were wrecked by the storms and in the fog of the northern Pacific. Many lives were lost.

Many others failed to arrive on the Fraser because the Nez Pierce Indians resented the great mass of miners crossing their territory, and because they were out for loot.

The miners coming up from California, hearing of the raids on the mining parties on the eastern side of the Cascades in Washington, waited at The Dalles until a force of sufficient strength had gathered to guarantee protection, before setting out across the Indian country. The Nez Pierce now confronted by a well-armed force of from 500 to 1,000 men hesitated to attack directly. They concentrated on night raids, not in hopes of taking any white scalps, but to steal, or at least to drive off the riding horses and the pack animals (there were no wagons). Once a fair percentage of these were gone, those on foot began to fall behind. The cavalcade broke up into small groups as the men on foot were unable to keep up with those still mounted. It was a case of keep up, or else; and "the devil take the hindmost."

There were two reasons for this. First, those still able to set a fast pace rushed headlong northward trying to beat the swarm of miners coming by boat. They were so innoculated by the gold fever that nothing would stop them from rushing ahead to get to the gold country before all the good claims were staked. Secondly, the constant Indian harassment had the miners so jittery that they did exactly what the Indians wanted. They broke rank and made a headlong dash for the Canadian border, where they had been led to believe the Canadian Mounties would provide needed protection.

Had the Indians had any kind of effective leadership, such as provided by Chief Joseph 20 years later, the great stream of miners would have been wiped out with the loss of thousands of lives. As it was, the warriors soon became satiated with scalps and loot. They withdrew to celebrate and to brag about their exploits. Even so, several hundred scalps were "lifted," and more than a hundred more men died of exhaustion before reaching the relative safety of the Fraser.

When they did arrive, they soon found that they would have been better off staying in California and working for wages or competing with the Chinese in working the poorer deposits. They would have recovered more gold and they would have been in less danger from the elements and from the Indians.

Later, in the 1880's and 1890's, rich deposits were found both east and west of the Fraser diggings, but the California economy was not affected any more than the Alaskan rush affected it a few years later. By this time the California mining industry had settled down to a few deep quartz mines and a number of drift mines tunnelling into the Tertiary gravel leads. So, while the know-how of the California miners was called on to develop the new fields, only a few technical men actually left the California mines, and these were replaced by a backlog of men coming out of the newly established School of Mines at Stanford and from the various State Universities.

During the 1860's, three major rushes developed within the United States; the first to eastern Oregon, then to Idaho along the Salmon and the Snake Rivers, and finally to Montana. In each of these the early arrivals got more than a few colors as had been the case on the Kern and on the Fraser. The techniques for gold recovery had been worked out in California. The first arrivals in these fields were able to clean up the rich gravel in a very short time. This, of course, led to exaggerated stories of the wealth of the diggings because, whereas it took a miner several years to wash the gravel from a bar in California, the same deposit would now be worked in a few months. This made it show a return which could only be matched by the richest of mines in the Sierra foothills.

As a result of the improved techniques, the valuable gravel was quickly exhausted and ghost towns developed. The quartz mines were likewise mined to the limits of the surface enrichment and then they too were abandoned. In a few cases, as with the copper mines of Butte and the lead-silver mines of Coeur d' Alene district in the panhandle of Idaho, the mineralized veins extended below the surface enriched area. Those mines there are still operating.

The rushes to Australia and the opening of the South African mines pulled men away from the California mines, but mostly they were engineering and technical workers and were replaced without serious damage to the economy of the West.

CHAPTER VIII

GOLD FROM THE ANCIENT CHANNELS

As soon as the miners began working the stream beds of the Sierra, some began speculating on the origin of the gold. Of course, mostly they just accepted the fact that the gold was there. They concentrated all their energies on getting it out with the least amount of effort. Once they had exhausted the easily obtained gold from the benches and the bars, they joined the rush to the next "bonanza." This was always reputed to be much richer than anything yet worked. Some of the more intelligent, however, noticed that most of the gold was found below the 4,000 foot level, and that the amount of gold in the streams did not taper off as they prospected upstream, but stopped suddenly. This was true in the channels of the main streams and in their tributaries.

After the discovery of the gold bearing quartz ledges in the Grass Valley area, it was assumed that the sudden appearance of the gold in the streams was due to the cutting and the gradual erosion of quartz ledges. This released the gold into the stream gravel. Some of the more observant, however, noticed that the gold, even when it appeared suddenly, was not rough and sharp as it would have been if only recently released from the quartz matrix, but was rounded and water worn. It showed all the characteristics of gold that had been transported over long distances by running water.

Further prospecting showed that it had come from gravel deposits high up on the canyon walls; it most definitely was alluvial gold. If these deposits were worked, the miners found that the gravel was not a bench left by the changing course of the river. Instead, if followed, it went straight into the canyon wall and continued on until it broke out into the adjoining canyon of the next stream to the north or to the south. It might, of course, be interrupted by a faulting of the rocks forming the mountain under which it was buried. Or, as was most often the case, a change in the direction of flow of the ancient stream stopped the deposition of the gravel. No trace was left to be followed.

The observant miner, and the scientists, when they arrived, observed that these "blue gravel" channels, as they were called, were composed of very old gravels. The rocks which made up most of the gravel showed much deeper weathering (oxidation) than those in the gravels of the modern streams. Also, it was very often cemented with lime and silicon. This cementing formed an almost solid conglomerate rock. But of more importance, it contained fossils which proved it to have belonged to the Eocene Geological Era.

In some cases the gold bearing gravel was not concentrated, but was scattered over large areas and was often hundreds of feet in thickness. These gravel fields showed a certain amount of assorting as if they had been lain down on the bottom of an ancient lake.

These lakes were probably formed by lava flows running across and damming the Eocene rivers. Lakes were formed, and the great, nearly flat, lake bottom deposits, such as those near Dutch Flat usually contained some gold. In that area the gravel was not washed away and the gold it carried deposited on, or near bedrock, but the entire load of the river was spread out over a wide area. In such a case the gravel contains only a few cents of gold in each cubic yard.

The terms Tertiary and Eocene will be used repeatedly. Actually the Eocene is the third from the last epoch of the Tertiary Age. Included in the Tertiary are the Paleocene, the Eocene, the Miocene, and the Pliocene, in that order. While millions of years were involved in the Tertiary, in reality it was the next most recent age in the geological history of the earth. It is followed by the Quartenary, which in turn is composed of the Pleistocene and the Recent. The latter being the present, or the Age of Man.

The Quartenary began with the uplifting of the Sierra Nevada Mountains, followed by the glaciation period. During the most recent of the glaciation periods, the Great Lakes were gouged out of the eastern and mid-western parts of the United States. Niagara Falls were formed, and the mountain valleys such as Yosemite and Hetch Hetchy were plucked out of the dioritic granite of the high Sierra range. In addition thousands of cirques [glacier cut depressions] were formed and later filled with water to become lakes such as Donner, Gold, and Mammoth. Hundreds of these still dot the mountains from south of Whitney to the Siskiyous.

These geological episodes in the history of the Sierra can be followed by those who are able to read the history of the earth in the rocks and in the gravel beds, by examining the fossils they contain.

For example, it is relatively easy to prove that the gravel from the hydraulic mines in the Red Dog — You Bet area was from the Eocene age. Litterly hundreds of acres of the outwash plains were, until recently, covered with petrified wood. The wood came from trees that grew only during that period. At one time during the late 1920's, it was the vogue to build fireplace facings out of this material. Hundreds of truck loads of the "wood" were hauled to southern California, and incorporated into the decorative scheme of the homes of the movie moguls of that period.

The trees from which the petrified wood came, grew in the lush river valleys where the Eocene streams were depositing the gold-bearing gravel. Suddenly great volcanic eruptions dammed the streams and submerged the forests in volcanic mud of very high silicon content. No oxygen could get to the wood to cause it to decay. The cellulose content was replaced, molecule by molecule, with silicon dioxide. The silicon compound now took the configuration of the original wood cells, and logs of stone were formed. These contained every detail of the original wood, even to the knots and branches. In the two decades between 1860 and 1880, the hydraulic monitors exposed the "wood" for our examination and for our use.

Between 1848 and 1910, the U.S. Geological Survey estimates that $1,200,000,000 in placer gold was taken from the stream gravels of California. Of this, $300,000,000 came directly from the Tertiary gravels, extracted by hydraulicing and by drift mining. Most of the remaining $900,000,000 must have come from the Eocene and later Tertiary streams. The modern streams have been tearing at the volcanic debris for a little more than one million years, hardly time for them to do more than wash out and re-sort the ancient dust and nuggets from the banks where they had been deposited as long as 50 million years before. Certainly the modern streams have not had time to do more than merely scratch the surfaces of the quartz ledges that have been re-exposed by the erosion of the modern streams.

The early miners did not understand the geology involved in the formation of the blue channel gravels; and most of them "could have cared less." They were only interested in getting out the gold. It was not until the completion of the Central Pacific Railroad, and the development of the science of Geology in the European and in some of the Eastern Universities of this country, that engineers with a geological background were able to visit the mines and to compile the data needed to show the real history of the ancient channel- and lake-deposited gold. In time the development of the Department of Geology at the University of California at Berkeley, followed by the College of Mining, produced the trained men to pick up and to carry on the work of Aggisez of Harvard, and others. Still later, the specialization of mining engineering at Stanford University contributed much to the real solution of the mystery of the Blue Channels.

At first, geologists believed that the area which now comprises the Sierra Nevada Mountain region of central California was much less mountainous than at present. Instead of sloping from east to west, the grade was from north to south. The rivers of the Teritary times followed the grade. By following the series of Blue Channels, it seemed apparent that the rivers had come from the north and had continued on south to just south of Yosemite, from there they turned west and combined into a great flood in the San Joaquin Valley and out through the Golden Gate into the Pacific. Some even speculated that this great river was indeed the Columbia River. They reasoned that the great stream was eventually dammed off by the lava flows which now comprise the Plateau regions of eastern Oregon. All evidence was, of course, lost under thousands of feet of volcanic deposits. Once dammed, the great river coming down out of Canada was forced to turn westward and cut its way to the sea at Astoria.

Some of the lesser informed miners still believe in the Columbian hypothesis. Later investigation, however, has shown that the Eocene deposits were made by smaller rivers running in approximately the same locations as do the present streams. At the end of the Eocene period, the old rivers were buried, the fault block which we now know as the Sierra Nevada tilted from east to west, and the modern streams began their work of tearing down the newly formed hills.

Under the conditions of great rainfall existing during the Tertiary, whole mountain ranges were eroded away. The gold bearing quartz rocks were broken away; the rocks were rolled along the stream bottoms, pounded and buffed and gradually reduced to sand. During the grinding process, the free gold they carried was released to become flakes of dust and to be rolled and rounded into nuggets.

Not all the quartz containing free gold was completely disintergrated. The early miners and sometimes the modern prospectors picked up rounded quartz rocks which showed specks of gold protruding from their sides. Often, when they are broken open, they are found to be filled with gold. These rocks are heavier than the ordinary gravel, so they lag behind as the material being washed is put through a sluice box. The experienced miner watches for these, as a single rock no bigger than a man's fist may contain several hundred dollar's worth of gold. Anyone finding such a quartz rock today is indeed fortunate as the jewelers in the foothill stores will pay a premium of fifty dollars an ounce, or even more, for the gold it contains. They use the entire rock for the manufacture of polished quartz jewelry. Highgrade, or the so called "jewelry rock" from the deep mines will not do for this purpose unless it has been salvaged in relatively large chunks. The shattering effect of the blasting to get it broken loose from the underground ledge ruins it for jewelry making. With a diamond rock saw, the gold bearing boulder is cut into thin slabs. Pieces are cut out and ground to fit into the settings of rings, lavalieres and other pieces of jewelry. The resulting quartz-gold pieces are in great demand by the tourists as they flock through the mining country.

For exactly how many millions of years the prehistoric rivers carved away at the gold bearing hills and sorted the gravel to make the rich blue gravel channels, we have no way of knowing. It must have been for a very long time. We do know that suddenly the whole process was stopped by the eruption of a series of volcanoes, extending from south of Mono Lake to Mt. Lassen. The volcanoes spewed out lava, and blasted out the ash which covered the whole area with deposits thousands of feet thick. Like the city of Pompeii being preserved for us by the rain of volcanic ash from Vesuvius, the gravels in the ancient stream channels were frozen in place by the sudden violent eruption of these ancient unknown volcanoes.

Millions of years passed. The volcanoes and much of their deposits of lava and ash were eroded away. Only the basalt caps of some of the higher mountains in the Sierra Nevada, and some of the more compacted and cemented deposits of the ash remain.

Just recently, as geological time goes, another great change in the earth's crust took place. The chain of mountains we now know as the Sierra Nevada was formed. These mountains were created by a great fault block tilting up on its eastern edge and down on the western half. This block is a large section of the earth's crust, extending some one hundred miles from east to west and running about four hundred miles from north to south. It extends from the foot of the escarpment in the Nevada desert to the east of the crest of the Sierra, to somewhere under the floor of the Sacramento and San Joaquin Valleys, and from Mt. Lassen in the north to the Tehachapis, to the south and east of Bakersfield.

That this fault block exists is evident to

anyone who has ever driven across any of the passes from the Great Central Valleys to the Nevada deserts. At Donner Pass, for example, a long gradual grade leads from Roseville, at near sea level, to the summit which is slightly over 7,000 feet.

Old Highway 30 then plunged down the face of a cliff to Donner Lake some 1,500 feet below. The road switched back and forth to reduce the grade, but in doing so it formed the worst bottleneck in the state, blocking the flow of traffic between northern California and the Reno-Tahoe resort area, as well as slowing the normal heavy transcontinental flow of cars and trucks.[1]

The break at the summit becomes so sharp as the mountains progress to the south that Tioga Pass in Yosemite Park is the last crossing of the Sierra until Walker Pass at the southern end of the range is reached. Even so, Ebbetts, Carson, and Sonora Passes are too high and the road too twisted for heavy traffic. In fact, they are not maintained for winter traffic. The problem of deep and drifting snow is much too great for even modern machinery to handle.

Motorists who live in the San Joaquin Valley

---

[1]Old Highway 30 has been replaced by Interstate 80. It crosses the Sierra one mile north of the Pass made famous by the Donner Party. The four-lane divided highway now angles down the north walls of the "cirque" at the foot of which was located the camp where, during the winter of 1846, the pioneers were forced to eat the flesh of their dead companions to survive. The old road climbed straight up the face of the granite walls at the head of the cirque. I-80 loses much of the spectacular view of the lake which the travelers on the old road enjoyed. On the present road, we bow to the "Great God Speed," but lose much of the enjoyment of the trip to Reno!

must make a choice during the winter months. Either they go south to Tehachapi, or they must go north to Placerville (Old Hangtown) to cross into Nevada. The other passes are closed by from twenty to sixty feet of snow. The railroads recognized this problem. There are no tracks across the mountains north of Tehachapi until Donner Pass, north of Lake Tahoe, is reached. Even here, mile after mile of snow sheds must be provided to keep the tracks free from snow during the winter.

It was this faulting and tilting that changed the drainage in the Sierra so that now the streams flow from east to west in almost straight lines. It also increased the average grade to nearly 200 feet per mile. This very steep grade led to an extremely rapid rate of erosion. Now every one of the major streams is running in a deep canyon, with great semi-bare walls exposed to the searchers for gold. Mostly, the rivers are now running through the original unaltered rock of the earth's crust which contains the quartz veins, which in turn, carry gold and other valuable metals.

The very rapid erosion carried on by the running waters of the mountain streams has also carried away a high percentage of the lavas and other mantle rocks which once covered the eastern portion of the fault block. Now the roads through the high passes pass through an area of granites, seynites and diorites. These are the gray rocks so familiar to those who know the high country along the John Muir Trail and the walls of Yosemite Valley.

The early prospectors observed that there was no gold in the gravels along a stream unless there was a high percentage of white quartz pebbles mixed with the sand. If, while he was prospecting a stream, he noticed that the gravel suddenly became much darker, no more gold would be found.

Naturally he started searching for the source of the quartz pebbles and for the gold which accompanied them.

Sometimes he found the quartz "stringers," "veins," or "ledges" which intruded into the deep seated igneous rock (country rock). If these deposits were not found, the prospector knew that the source of the white stones must have been the deposits of stream gravel laid down by Eocene Rivers. Once that fact was established, the miner searched the areas high on the canyon walls above the country rock in sedimentary or volcanic deposits. It was there that the Eocene gravels were to be found, and if found, most usually produced rich diggings.

Once a channel was found, the miner was still in a quandary as to how to proceed to get the gold. Water was always the major problem. It could be miles to the nearest supply. At first the gravel was mined just as coal was mined in the east. That is, the channel was followed with a tunnel and the semi-cemented auriferious gravel quarried out. Once the gravel was delivered at the tunnel mouth, it still had to be carried to the river below. If the miner was prosperous he used a string of pack mules. If the topography was right, a road might be built and gravel hauled by wagon. In any case the gravel must be very rich to make it pay.

As in the case of the bench gravels, the miners reasoned, "If the gravel can't go to the water, the water will have to go to the gravel."

So diversion dams were built, ditches were dug, flumes were constructed and water was delivered to the sites where needed. The procedure was exactly the same as for the bench gravel found at lower elevations, and much closer to the modern rivers. The channel gravel might be a thousand feet or more up the canyon wall; in which case, much longer ditches were needed to deliver the water to the higher elevations.

As noted before, the rivers had an average fall of about 200 feet per mile. To keep the water flowing rapidly, the ditches needed a fall of about 25 feet per mile; so there was a net gain of 175 feet per mile of river. Therefore, about 25 miles of ditch and flume were needed for each 1,000 feet of elevation gained. Of course this would be true only if the canyon walls were exactly parallel with the bed of the river at all times. Because there were side canyons to be crossed by following up one side and down the other, a minimum of double the milage was always needed.

As a result, hundreds of miles of ditches were dug during the 1850's and 1860's. Again the construction and maintenance of the ditches was much too much for a single miner, or even for a group of miners. Consequently, larger companies, some with capitalizations of hundreds of thousands of dollars were formed. Shares were sold to the general public. As a result, the major stock holders were often not mine owners at all, but were merchants, bankers, and trades people looking for high profits from the investment of their surplus funds.

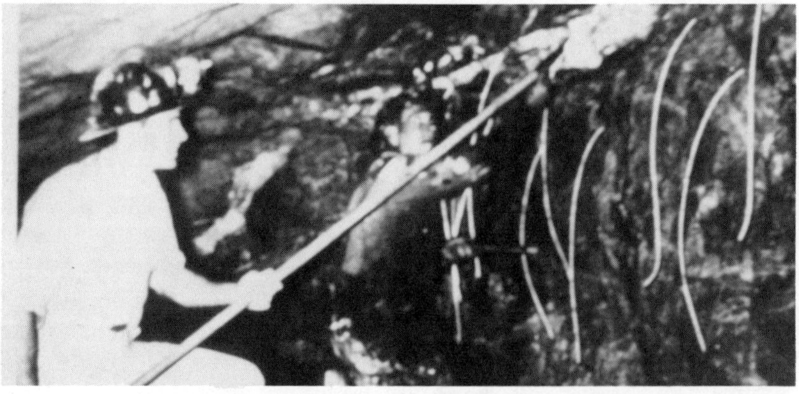

TAMPING HOME POWDER UNDER-CUTTING LEDGE

CHAPTER IX

THE HYDRAULIC MINES

Four short years after James Marshall made his discovery on the American, Joe Wood, a preacher's son from Ohio, and John Payne, a carpenter from Maine, were ground sluicing on the Yuba River, ten miles upriver from Marysville. To get water from the creek, coming in from the north side of the river to their diggings, they built a wooden flume. The grade of the flume was so great that when the direction of the flume had to be changed to get the water to a spot not as yet washed, the rush of the water caused it to shoot up over the side of the flume and be lost. To stop the loss, Payne nailed another board across the top of the flume. When it became necessary to change the direction of the flume once again, they did not build more flume; instead, they hit upon the idea of fitting a piece of cloth around the square end of the flume. The sides of the cloth were then sewn together forming a sort of hose. The pressure of the water in the box, plus the added pressure built up by the fall in the hose, caused the water to rush out of the end with sufficient force to help materially in the breaking up of the partially cemented gravel.

Both Wood and Payne had a hankering for the gold they knew was in the gravel they were working, but they also had an aversion to hard work. This, along with the incredible luck in having just gone to town to buy some muslin to make into some much needed shirts, led to the development of the hydraulic hose. Cloth of any kind was in very short supply anywhere in California in 1852. The store in Marysville had just sold its last bit of cloth, except a bolt of sail cloth that the owners had picked up during their last buying trip to San Francisco. That particular type of material was available because every ship that came into the harbor immediately lost all its crew. The Captain would beach it and sell off all the ship's stores. He then took off for the hills himself. Payne and Wood were told that this was the last cloth of any kind in town. "Take it or leave it!" They took!

The sail cloth had been made to withstand constant wetting, and to hold against high winds without damage. It was exactly what was needed to make a hose of considerable strength. The partners then found that they could taper the lower end of the hose and increase the pressure of the water. As the height of the trapped water in their enclosed flume and in their canvas hose increased, the water pressure increased in direct proportion to the head, or the height of the column of the enclosed water.

Next they were able to buy a short length of leather fire hose, with a brass nozzle attached, from a volunteer fire company in Yuba City. They added this to the end of their canvas hose and found that it was of great aid in directing the rushing water to the exact spot where it would be of most use. This was the first hydraulic nozzle. To withstand more pressure, they made their hose of several thicknesses of canvas, then increased its strength still more by wrapping it with strips of rawhide.

Every miner going up river stopped to see the new contraption. It was an immediate success. Almost to the man, they rushed back to Marysville and Yuba City to buy the supplies they would need to copy the new labor-saving device. Naturally the towns were stripped of everything which could be of use by the miners. Sacramento, and then San Francisco, soon felt the pressure for the needed items. Many months were to pass before orders could be sent to the east coast or to England and the supplies sent by clipper ship around the Horn. Once they did arrive, the mining industry in California was revolutionized.

It is not a matter of record if Anthony Chabot, a Nevada City miner, who had been removing the overburden and scraping the bedrock on the hillside north of town, saw or heard of the operations on the lower Yuba, or, as so often is the case, a much needed innovation was worked out simultaneously by two or more men. In any case, Chabot carried water to his claim through a wooden flume from a miner's ditch on the hillside above. To get the water to the spot where it would do the most good in carrying away the ground cover, he cut the leg off a discarded pair of pants. The larger upper end he nailed to the open end of his flume. Shortly, he also found that by pinching down the delivery end of his hose, he gained pressure and more efficiency.

Chabot normally gets the credit for having invented hydraulic mining because his was the first attempt to move a relatively thick overburden to get at the gold on the bedrock below. The Wood-Payne mine had been on gravel already exposed by the running river water. The practice of hydraulicing open gravel deposits soon died out. Chabot's method of mining, however, expanded rapidly, and became the main source of gold coming out of the California hills.[1]

"Mat" Matheson, another placer miner, saw Chabot's mining method. He had a better idea. He took the stove pipe from the stove in his cabin

---

[1]Chabot used his new method of mining to become quite wealthy. He migrated to the Bay area and invested his wealth in land. He donated Chabot Observatory, and park land to the City of Oakland.

and several sections which he was able to buy from the local store. These he fitted together to form an iron pipe more than 10 feet long. He reversed the process of working his claim. Instead of introducing the water at the upper end of the ground he was working and allowing the water to loosen and carry away the overburden, he ran his pipe to the lower edge of his claim and used the additional pressure gained by the use of the stronger pipe, to blast the unwanted material along with the gold-bearing materials, into a sluice box.[2]

Once again there was a rush to the sources of supply. Every blacksmith in the mining country was immediately put to work fabricating pipe out of anything which could be made into a tube, or could be made into a nozzle to fit the canvas hose which still was used to deliver the water under pressure to the desired spots.

To get more pressure, to blast down larger and harder cliffs above the gravel deposits, longer and longer pipes were used. Within a short time the canvas hose had to be enclosed in tight-fitting iron rings to keep it from bursting. Now the problem became one of holding the nozzle on target. The back-lash of the nozzle became so great that, first two, then three men were required to hold it. Even these were not always enough. The hose would break away and start lashing out like a great wounded serpent. Eventually the whiplash became so great that not only would arms and legs be broken, but men

---

[2]Some authorities state that Matheson was ground sluicing just below the Eureka Mine in Grass Valley. When he depleted the rich ground there, he moved his operations, along with his stove pipe. His method was adopted by so many that he is often credited with being the "father of hydraulic mining."

were thrown against the gravel banks with such force that every bone in their body would be broken as their life was crushed out.

Within two years a blacksmith by the name of Fisher invented a swivel joint for connecting the iron pipe directly to the heavy nozzle. This made the hose obsolete. Again the pressure was increased, and still more men were added to hold the "monitor" or "little giant" as it was sometimes called. They, too, sometimes failed. The heavy nozzle would break away with disastrous results. Finally, one of the ingenious Yankees from off a sailing ship, got tired of seeing men thrown against the cliffs or smashed by the blast of the water as it rushed out of the nozzle when it could no longer be controlled by the "jet men." He solved the problem by counter-balancing the nozzle by attaching a box filled with a ton or so of rock to an eight-foot boom affixed to and pointing in the opposite direction as the nozzle. Now a single man could control the monster with one hand, as it blasted down the gravel banks and washed them into the sluice boxes below.

The ditches dug for the bringing of water to the bars and benches to be used in the Long Toms and in the sluice boxes, faded into insignificance when compared to those needed to bring the water to the batteries of monitors as they ate into the Eocene river gravels.

Today, the motorist driving over I-80, parallels the hydraulice cliffs at Dutch Flat, some ten miles east of Colfax. A sweep of the eye shows that thousands of acres of relatively flat deposits were washed away during the three decades the monitors played against the hundred foot cliffs.[3]

---

[3]The State now maintains a "rest stop" for west bound traffic at the foot of the cliffs which were left standing, and still stand at the edge of

The Cherokee Mine on the San Juan Ridge, alone, shot 36,000,000 gallons of water a day at the cliffs during the dry season. During the winter and spring when more was available, twice as much water was used. In the spring of 1873, the Hungarian Mine on the Yuba River began operating. It had a 300 foot dam on the river above to store water for the dry season. Their four monitors worked under a hydrostatic head of 300 feet, each throwing 1,000 miner's inches of water against the cliffs with such force that 500 pound boulders were thrown 50 feet into the air, and acted as battering rams to help break up the compacted gravel as they smashed back to earth.

With literally hundreds of such monitors going, it is of little wonder that things began to happen in the valley below. The great torrents of water succeeded in carrying over two billion cubic yards of "slickings" (sand, gravel, and mud) into the Sacramento Valley during the thirty years of peak operation. This would cover an area 30 miles long and 30 miles wide, to a depth of 30 feet! Unfortunately for some, it was not spread out over the 900 square miles.

Charles Justis of Wheatland, near the mouth of the Bear River, testified to orchards being completely covered by the great "octopus" of piling gravel. W.H. Drum of Yuba River testified in court, in 1875, that his two-story house was buried to the eaves in sand and gravel, and that when he had built the house in 1853, its foundation was 23 feet above the high water mark.

In 1878, the State Mining Debris Commission took testimony to the effect that where the river boats once docked at Marysville to discharge cargo bound for the Northern Mines, now rowboats scraped

---

[footnote 3, con't.] the highway. A turnout parking area has also been provided for east bound traffic.

bottom most of the year. Where $1,000 an acre orchards once stood, now only barren sand and gravel banks were visible!

The traveler today, when crossing the levees, can readily see that the streets of Marysville are below the level of the water in both the Yuba and the Feather Rivers even during the dry season.

It has often been stated, and never refuted, that the loss to the ranchers and the cost to the government and to the state in opening and in keeping open the channels of the streams in the Sacramento Valley, has more than offset the value of all the gold recovered by the hydraulicers during their 30 years of uninterrupted operations. And the end is not as yet in sight. As late as 1955, the great Christmas flood crumbled the west levee of the Feather River below Marysville. The escaping water destroyed thousands of acres of crop land and orchards. It also trapped several hundred families at the cost of thirty-eight lives! Again in 1964, only the timely completion of the Oroville and the Folsom dams saved the valley from yet another disaster.

Almost as soon as the piles of slickings began piling up in the Sacramento Valley, the ranchers began appealing to Sacramento for relief. They got it; but only after a long and bitter fight, and after the Federal Courts had been drawn into the controversy, to rule in their favor.

This did not mean that the monitors were silenced. There was too much money involved to let the rulings of one old man in San Francisco interfere with their task of tearing down mountains and depositing them on some stupid farmer's land. The mining lobby in Sacramento saw to it that there were loop-holes in the law. These were put there to allow the mines to keep the profits rolling in.

One of the loop-holes engineered into the law by the lobbyists stated that the mine inspectors must actually see the monitors in action before the illegal miners could be brought into court. Every possible obstacle was thrown in the way of those trying to stop the hydraulicing. Mine inspectors were hired, but it was impossible to get the Legislature in Sacramento to appropriate the money needed to pay the travel expenses of the men charged with the enforcement of the law; so, in effect, they were chained to their desks in the capital city. If they started for the hills paying their own expenses, they were preceded by "pony express" riders mounted on the best horses in the country, dashing ahead like Paul Revere, shouting "the inspectors are coming, the inspectors are coming!" When the inspector did arrive, the sluice boxes and the streams could still be full of slickings, but the monitors were silent; nothing could be done.

Every stranger coming into the mining country had better be able to prove his identity, and have a good reason for being there, or a committee of rough characters would escort him to the edge of town and send him on his way with a warning to "keep going, or else!"

Two years after the mines were supposed to have been closed, Tom McGuire from Grass Valley was hired to teach the one-room school at the Malakoff Diggings. On the first Saturday after his arrival, he was introduced to the intricacies of the operations of a monitor. He was hired to handle the "swing bar" for one-half day only. But that was enough. By accepting the two dollars, the going rate at the time, he became just as guilty of breaking the law as were the mine owners. It was now safe to let him out of the hydraulic mining country. If he talked, he too could go to jail![4]

---

[4]Tom McGuire died in 1968, at the ripe old age of 93. Without a doubt he had been the last

The first commercial, long distance telephone lines in the world were installed by the hydraulic mine owners. First they were used to allow the men at the mines to control the flow of the water by telephoning the gate keepers at Lake Spaulding and at other mountain reservoirs, advising them as to the amount of water needed. This worked so well that a line was built from the road house at the junction of the Marysville-Grass Valley road with the mountain road which crossed the Yuba River at Bridgeport. The Miners' Association kept a man at the road house. His duty was to watch every man taking the road to the mountains and call ahead with a warning if anyone not known to be friendly to their cause took the turn off.

Eventually, of course, the law prevailed and the blasting giants were directed against the Pelton Water Wheels to convert their snarling, hissing jets into a tamed flow of electrons over the copper wires to light the cities and to run the motors of the industries of northern and central California.

What of the future of hydraulic mining? It is known that at the present price of gold, there is well over $2,500,000,000 of recoverable gold locked in the gravel banks of Nevada County alone![5] But locked in, is the correct term. Hydraulic mining is the only possible way by

---

[footnote 4, con't] living man to have worked for the hydraulic mines during the last years of their operations. He was the son of a miner. He had lived all his life in Grass Valley. It was not until he had attended the University where he got his degree in engineering that he came to realize just how wrong he had been.

[5]Multiply that amount by 15 to get the value in the 1980's, when the world price of gold averaged $450.00 per troy ounce.

which that tremendous amount of wealth can be recovered.

During the last years of the Depression, the Government built the Englebright Dam on the Yuba River just south of Timbucktoo. This was to allow the hydraulic mines to start up once again. The slickings would be trapped behind the dam and would not harm the ranch lands below. However, before the miles of pipes could be installed, the Japanese bombed Pearl Harbor. The pipe was immediately requisitioned by the defense industries. After the War, the cost of mining plus the cost of storing the tailings back of the dam became so great that no one could see a profit to be made by opening the mines. So the mines remained closed. Even at the present high price of gold, the value of the water for the generating of electricity, and for irrigation is so great that the hydraulic mines still would not show a profit sufficient to warrant the millions of dollars needed to begin their operations.

Reopened main shaft. Empire Mine.

CHAPTER X

SLUICE BOXES AND THE CHINESE

Once Chabot, Matheson, and the others had perfected the hydraulic method of recovering gold from lake beds and ancient river channels, the need for miles of ditches, flumes, dams, and pipe lines forced mining in California to become big business. Stock companies were formed, often with the major stock holders living in the banking centers of the world, especially in San Francisco. With the advent of absentee ownership, the moral code of the miners in the hills changed abruptly.

While the miners were working as individuals, and even after small groups began working together, it was common practice for the workers to go to town, or on a short hunting trip with no thought of securing their exposed wealth. Their cabins were left open and unlocked. Their sluice boxes and Long Toms were left with the results of days of washing the richest of gravel lying back of the riffles, and their mining tools and other personal belongings, left in the open with complete confidence that they would be exactly where left, upon their return.

With rich men living in the cities owning the mines, the tradition that no one would ever touch another man's gold became a thing of the past. With miles of sluice boxes strung along the sides of isolated canyons, the temptation became too great for those who wished to live well, but lacked the character to do the hard work required to accumulate the necessary gold. The mining companies posted guards along the boxes where the amount of gold was apt to be the greatest, but all too often the guards themselves reasoned that a small addition to their meager wage was more or less legitimate.

Even though there might be but little gold trapped back of the riffles where no guard was posted, the highgrader always got enough value to make his actions worth the very small risk he took by robbing the box.

Mercury, or quicksilver, as miners called the shiny liquid metal, was always put behind the wooden cleats forming the riffles in the sluice boxes. This was to aid in the separating of the gold from the other heavy materials in the rushing water below the monitors. Mercury has a density of about thirteen. Gold is nineteen times as heavy as water, while most sands and gravels have densities of two and one-half to five. It is readily seen that water, pebbles, and sand will float over the surface of the mercury, but gold will be submerged and trapped.

Not only did the gold sink into the mercury, it was also amalgamated by it. Amalgamation is a special kind of solution where one metal dissolves in another. The amalgam, being still heavier than the mercury, stayed at the bottom of the pool behind each of the sluice box cleats. The sluice-box robber, using a small flat-point shovel, reached into the sluice boxes to scoop out the mercury and the amalgam. If no gold happened to be trapped at that spot, the robber's efforts were not entirely wasted. The mercury was worth about a dollar a pound, and it took a surprisingly small amount to weigh a pound.

How much gold was actually highgraded out of the hydraulic sluice boxes, will, of course, never be known. It was enough, however, to cause the Malakoff Diggings at North Columbia to drive a tunnel through a ridge from the bottom of its hydraulic pit, to the branch of the Yuba River which was destined to carry the slickings to the valley below. The normal procedure would have been to build a sluice box around the shoulder of the ridge. This would have accomplished two things: carry the washed-out materials from gravel cliffs to a point of disposal, and recover the gold during that process. Inside the tunnel, the rough bedrock of the floor acted as a mile and a half of effective riffles, and it was easy to guard both entrances to keep the highgraders completely away.

Gold was the common medium of exchange during the time the hydraulic mines were in operation. There was never any trouble getting rid of the gold once the robber had it in his hands. It could be dust or "sponge cake," or "sponge gold" (the gold that remained after the mercury had been distilled from the amalgam). The local grocer was used to receiving it in payment for supplies, or he was prepared to buy it outright at $2.00 per ounce under the price he would receive for it at the Mint in San Francisco. During the three decades that the mines operated at capacity, an average of slightly more than $50,000,000 per year was shipped to the Mint. With that much gold collecting in the sluice boxes, it is not unreasonable to believe that $1,000,000 per year was highgraded.

It is also quite probable that the gold from some of the early "secret mines" were the results of successful raids on the exposed boxes. With new mines being found with great regularity and with claim jumpers standing by to move in on an unprotected claim, it was common practice to keep the location of a new mine secret until it could be properly protected. What would be more simple than for a man to rob a sluice box, disappear into the hills for two or more months, then reappear with the story that his sudden wealth came from a rich strike he had made during a prospecting trip?

When the monthly clean-up below the monitors at any mine showed far less gold than average, the management of that mine was almost sure that the boxes had been robbed. As always a scapegoat had to be found. In most cases it was the Chinese who came under suspicion.

Thousands of Orientals had been imported to do the drilling, blasting, and grading during the building of the Central Pacific Railroad (now the Southern Pacific) across the Sierra. They pushed tracks on across Nevada and Utah to Promontory Point where the golden spike was driven. That done, most of the construction workers were discharged. They could not be expected to return to China, at least not until they were returned in a pine box to be interred with their illustrious ancestors.

Meantime, they had to eat. They were not welcome in the industries of the state, nor for the most part, by the ranchers. Great numbers drifted into mining, but in most of the mining districts they were not allowed to take up claims; or if they did succeed in locating a bench or a bar in an area that had been overlooked by others, it was a simple matter for an American to jump the claim. The Oriental had no standing in the courts, so the claim jumper simply moved onto the property and began to mine. The "Chink" could only move on and hope to find another place to make a dollar or so a day.

After a few experiences of losing their good diggings, the Orientals found that it was more profitable to limit their mining operations to worked-over ground. There with infinite industry and patience they eked out a living by reworking the tailings left by the earlier miners, and ground thought to be too poor to be of any value. If by a chance a Chinese miner did make a lucky strike, he was told to keep very still about it, and to send the gold to his Tong in San Francisco rather than to the Mint. Not until many years later did it become known that fortunes had been secretly smuggled out of the mining country.

Very close to a million ounces must have been scraped off the bedrock under the public buildings standing on the gravel at Scott Bar in Siskiyou County. Who can know how many more millions were taken from the surface, the quartz ledges, and the pockets from Coarse Gold in Fresno County to Happy Camp on the Klamath?

In spite of all his handicaps, the Chinese miner lived, and often prospered, much to the consternation of the ne'er-do-well white fortune hunter. So what was more natural than to blame everything that went wrong on the Chinese? If the trail to the real sluice-box robber got too hot, a sure method of throwing the law-man off the scent was to start the rumor that "Ho Pong" was seen near the boxes. A search would always show some fresh amalgam in his cabin (planted). This usually ended in a hanging, but strangely the sluice boxes continued to be robbed. This pattern was followed innumerable times until mining no longer provided a livelihood for the Chinese and all but a few gardeners, laundrymen, and green grocers had moved to the Chinatowns of Sacramento and San Francisco.

The Chinese were not only blamed for the sluice box robberies, but for everything else that went wrong in the mining country and in any other area where the Orientals lived and worked. This was only natural as they represented a minority group which was a real or an implied threat to the wage structure of the non-unionized laborers throughout the whole of California. Charlie's mule was indicative of the psychology of the times.

Charley Orr was mining a small Eocene gravel deposit which he had found on the south slopes of the Yuba River Canyon. He owned an old gray mule, Maud. He used her to pack the gravel down to the water below. He was not only fond of the old jenny, but he was entirely dependant for his living on the gravel she packed for him.

One Saturday night, Charley got back from having a few drinks at the local bar to find the gate down and Maud gone. Early the next morning he started looking for his mule and spreading the word up and down the river. One group was formed to go with Charley up stream, while another group followed the trail downstream.

After a couple of hours of looking and pausing at each bar as they passed to spread the word and to sharpen their eyesight with a shot of rye or with local corn whiskey, someone in the downstream party spotted a "Chinaman" leading a gray mule along a trail half way up the canyon wall.

With a "whoop and a hollar" they scrambled up the hillside to grab the unfortunate Foo Wong. Then they hustled him and the mule down to the settlement below. Miners' Court was immediately convened, and a trial held without delay. Someone suggested that they ought to wait for Charley to identify his mule, but that might take hours. They were in no mood to postpone "justice" for the thieving Chinaman who was caught in the act. It was the unanimous decision of the court that they take him out and hang him from the nearest oak limb.

Two hours later Charlie arrived on the scene to retrieve his mule. A quick inspection, however, revealed the startling fact that this was not his mule at all! It was not even of the right sex! The "Chinaman" had been right all the time in his protestations that the mule was his; bought and paid for. Another miners' meeting was called at once to consider what to do about the slight mistake. After due deliberation, it was decided that the joke was on the "Chinaman." Everyone had a good laugh and the meeting was adjourned to the nearest bar for another round of drinks before going back to their diggings. Of course, the miner who had first seen the "Chinaman's" mule stood for the drinks. It was only logical that he should. He had just inherited the now ownerless animal.

Actions such as these may sound heartless and callous to us, but such was the tenor of the times. Many other mistakes of a similar nature were made, and it was not always the foreigners or the Indians who were victims of the Kangaroo Courts. The miners were in no mood to spend needless time complying with the niceties of court-room procedures. A crime had been committed; the evidence, on the surface at least, pointed to the defendant. There was gold to be dug. Why waste time? "Let's get this over with," was too often the feeling; so the philosophy of the court was, "better to hang a few innocent men than to let one guilty man go free due to a technicality." Today our reasoning is: "better a hundred guilty men go free than one innocent man rot in jail, or pay with his life for the crime of another!"

This is not to say that the Chinese were always innocent when sluice boxes were robbed. They certainly did help themselves whenever they had the chance, but they were no more guilty than the white miners and, most probably, did not get a full share of the loot. But if the suspect's eyes were slanted, and he had been caught near a sluice box, he had just issued an invitation to a hanging—his own! The Caucasian miner, caught under the same circumstances, had a much better chance of being protected by the Anglo-Saxon concept of the law which considers a man innocent until proven guilty. In any case, the mine owners were more apt to listen to the white miner's "plausible" reason for having been in the vicinity of the sluice boxes.

As mentioned before, the sluice-box robber had little trouble disposing of his loot. This was true whether he scooped up dust and nuggets from back of the riffle cleats, or if it was mercury or gold amalgam. The raw mercury was easily separated from the amalgam by squeezing it in a wet chamois skin bag. The liquid mercury passed through the tiny holes which had been the original pores of the skin. The amalgam, and any chunks of gold, remained "in the bag" (this is most likely the origin of that cliche) in the form of a semi-solid lump. This was then heated in a retort over a charcoal fire. The mercury evaporated, was condensed and saved. The gold remained in the retort as "sponge gold." It was semi-porous like a sponge. This was then sold to the local gold buyer, the grocer, or the saloon keeper.

The price paid was, of course, dependant on the amount of highgrading going on at the time, and to how much risk the buyer was willing to take by buying from a stranger. For the most part, possession was all the proof needed to establish ownership. A known loafer, or the town ne'er-do-well might have to go down to Sacramento to find a buyer, who if questioned, would swear that he had known the seller to be a legitimate miner, and had every right to sell his own sponge gold. No farther proof was necessary. No gold buying license was needed, and dust, or sponge gold needed no more proof of ownership than did gold coins.

Later, after gold was nationalized in 1933, and a buyer needed a license, a miner coming into town from the hills had only to sign an affidavit that he had washed the gold out of the gravel on such and such creek, about so many miles above its intersection with "X" River. With this, the gold buyer was in the clear even though he might have known that the "bum" had not worked a day in twenty years.

Naturally, a few unlucky riffle robbers got

caught and went to jail. That is, they went to jail if they were American or European. If they were unlucky enough to have been born with slanting eyes, or with the brown pigment of the Mexican or of the Indian in their skin, they were fortunate indeed to escape a quick hanging.

Most of those caught and convicted were nabbed in the act of scooping the values out of the boxes. This was hard to do as the robbers almost always worked at night, usually during the dark of the moon, and preferably during a heavy storm. If they were surprised by a guard, they could escape into the darkness; and only if they ran directly into the arms of another guard, or fell in the darkness and broke a leg, were they apprehended.

During the heydays of the hydraulic mining in Nevada County, the robberies of the sluice boxes got so bad that the Blue Tent Mine hired extra guards for their boxes. As an added incentive, they posted a $20.00 bonus for every "Chinaman" caught robbing their boxes. Of course by "caught" they meant killed as the guards were told to shoot first and ask questions later. No questions were ever asked. If an Oriental was within gun shot of the box when the guard came along, he had just signed his own death warrant!

A peculiar thing happened. The robberies of the boxes stopped, but the guards continued to collect their bounty money! Every few days there would be gun fire along boxes well away from places where the "hunter" might be observed, and sure enough, there would be another dead "Chinaman," wet gravel still clutched in his hand.

At last the mine manager got suspicious. He had the doctor come out from Nevada City to inspect the corpse when the guard came in to report yet another Chinese dead, and to collect his $20.00. The doctor pronounced the Chinaman dead all right. In fact, he swore that in his professional judgement the man had been dead for some forty-eight hours. If they doubted his word, just keep the "stiff" out in the hot sun of mid-summer for another twenty-four hours; then they would have plenty of physical evidence to prove his contentions.

When pressed, the guard admitted that he had needed an extra twenty dollars. He had gone down to the creek where the Chinese were working the mine tailings; shot the nearest "Chinaman" and dragged him back. He made a serious mistake—he stopped at Higging's bar for a drink. One led to another, and before he knew it, he was two days late in getting back on the job. Anyway, he could not see what the fuss was all about. This had been going on for several months. The "Chinaman" would rob the boxes later; he was just stopping trouble before it started. This was his idea of a preventative war!

Even when caught in the act, the American sluice-box robber had an excellent chance of escaping a jail sentence. He hired a good lawyer and demanded a jury trial. It was "well nigh" impossible to find twelve men, one of whom had not done a little box dipping himself. The clever attorney knew this, and being so informed, he knew that all he had to do was to cast a reasonable doubt as to the reliability of the prosecution witnesses, and a hung jury was assured. Retrials were seldom granted. Mostly the District Attorney did not ask for one. He knew the "facts of life" in the mining country. If the case had been brought to trial, the public knew that there must have been a strong case against the defendant. That knowledge alone "black-listed" him. He would never be able to get employment around the mines again. That in itself was punishment enough in the eyes of most of the population.

Again, this was true only if the color of the accused man's skin was "right." All others could only pray for the opportunity to be tried in a real court, and hope for a non-hanging judge.

The Chinese sluice-box robber had one thing in his favor. He had a ready-made market for his gold. No questions would be asked, and none of the local gold buyers would have any knowledge of the transaction. A thriving Chinatown developed in both Sacramento and San Francisco. There, the Chinese were able to survive and prosper in spite of the relentless persecution. They did this by organizing into secret societies known as "Tongs."

The Tongs were groups of Chinese, organized along family lines for protection from the Americans and from other Tongs. During the construction days in the Sierra and in the deserts of Nevada and Utah, the Chinese learned by bitter experience that they could not expect a fair trial in the American courts. This was true whether the alleged crime was against a Caucasian or against one of their own.

So the Tongs set up their own system of dispensing justice. They made the arrests, conducted the trials, and inflicted the punishment, often to the extent of the execution of the prisoner. All this was according to their own code of justice. This was a far cry from the criminal code which was supposed to be in force in the State of California.

Often the suspect could not be brought to trial; so he was tried in absentia, and if found guilty, was sentenced. If the death sentence was imposed, he was hunted down by a "court" appointed executioner. Usually he was dispatched by the use of a knife or a hatchet. (It made far less noise that way, and thus the term "hatchet man" was coined.) Often the executioner was then tried in absentia by the offended Tong. So the "Tong wars" began and continued.

This taking of the law into their hands was, of course, much to the dislike of the city police, and of the State law enforcing agencies. There were, and are, exceptions; but it was into the twentieth century before the Chinese became intergrated sufficiently to understand and to trust, in part at least, the American system of justice. This came about after the enforcement of the attendance of all children in an American school, and the recognition that all children, even those with a tinted skin and almond shaped eyes were American citizens. Only when the Chinatown police detail could walk into Tong headquarters to demand and receive the cooperation of the Tong rulers, could the vicious circle of killings in retribution be broken.

During the mining days, every Chinese, even those working in the most remote mining country belonged to a Tong. As soon as a few ounces of gold were accumulated, it was sent by Wells Fargo, or by a trusted messenger, to the proper Tong headquarters. There it was converted into cash and invested in Tong enterprises. The headmen of the Tongs acted the same as the Board of Directors of a corporation. They regularly handled large amounts of legitimately mined gold; so it was no problem for them to dispose of a few ounces of stolen sluice-box dust.

The records of the Tongs in safeguarding the funds entrusted to them were far better than the records of the banks and the trust companies of the period. The Tongs invested their funds in everything from California ranch lands, with the title vested in the names of children born in the U.S., to banks in old China.

There are several descendants of the Tongs, now called the "Families" still operating in the Chinatowns of Sacramento and San Francisco.[1]

---

[1]The writer was for some years a real estate broker in the San Francisco Bay Area. At one time he had an occasion to enquire into the ownership

[footnote con't] of a piece of very valuable Bay shore property. He found it to belong to a certain Wong family, with an address on Grant Street in San Francisco. A trip across the Bay to that address, proved it to be a small Chinese grocery store with the usual pressed duck, dried octopus, and shark's fins, along with all the odors associated with imported Oriental foods.

The proprietor was located. He spoke excellent English. And why not? He was a fourth generation Chinese-American with a masters degree in English Literature from the University of California at Berkeley. He advised that a trip across the Bay was not necessary. Next time, please contact him at his home in Berkeley. This turned out to be in one of Berkeley's most prestigious residential sections. His home fitted perfectly with its neighbors, none of which would be assessed for less than a quarter of a million dollars today.

The Tongs apparently are still paying off!

MONITORS PLAYING. MALAKOFF DIGGINGS

Battery Box — Screen removed. Note renewable heads.

CAMS LIFTING STAMPS. 5 STAMP MILL

DREDGER TAILINGS, AMERICAN RIVER, FOLSOM

GENERAL BEALE. USAF PHOTO. CIRCA 1880.

Gold Dredge at Natomas.

CHAPTER XI

THE BIG TIME HIGHGRADERS

"Thou shall not take from thy cabin mate, nor hoist his gold dust to add to thine, lest he find thee out and straightway call his fellow miners together, and then hang thee . . . or give thee fifty lashes and two hours to leave the country, or brand thee like a horse thief with an "R" upon thy cheek to be known and read of all men." [The miner's eighth commandment.]

That commandment was taken seriously by the miners. It was usually perfectly safe for a miner to leave his cabin open and unlocked. Any stranger coming by was welcome, and was expected to use the cabin for shelter. He could use any food which might be available, but under no circumstances was he to take anything other than food. He was to leave the cabin clean, especially the dishes. Firewood must be replaced, and if the cabin owner should happen to leave his poke of dust lying around, he could expect to find it intact when he returned.

That the above did not apply to the miners in their relationship with the large mining companies is illustrated by the Jim Marshall story.

Jim was not related to the discoverer of gold at Coloma. In fact he did not arrive in the gold fields until rail transportation became available. By that time, of course, the easy pickings were gone. Fortunes were still to be made in mining, but it took capital to start the hydraulic monitors blasting at the gravel banks, or to drive the shafts looking for the paying quartz ledges. Jim did not have the capital, nor did he have the ability to organize a company which could raise the money needed to begin mining with any prospect of success. Like so many others who had come to California to make a quick fortune, he was forced to go to work as a laborer in order to eat.

Jim found a job with the Eagle Mining Company near French Corral. At first he worked as a mucker, raking the larger boulders out of the head of the sluice boxes to keep them from blocking the flow or crushing the wooden cleats which formed the riffles. The wages were two-fifty a day and "found" (room and board). This was not much, but it was better than he had made in Chicago driving a horse car. Certainly it was not enough to make even a beginning towards the fortune he had dreamed of as he had ridden the passenger coach across the prairie lands and over the deserts and the mountains on the way to the land of his dreams.

In due time, he was promoted to the swing bar of the hydraulic monitor. This was a responsible job. It took a certain amount of skill, and exceptionally good judgment, to keep the water playing against the gravel banks to carry away the overburden, and then to wash the pay dirt into the boxes where the gold could be trapped. Eventually he was included in the clean-up crew whose job it was to scoop the amalgam from the riffles, squeeze out the excess mercury, then retort the amalgam and replace the mercury in the sluice boxes. The sponge gold he then helped melt and cast into bars for shipment to the Mint.

By the time Jim had worked himself up to this rather responsible position, his wage had been increased to five dollars per day. This was more than most men made at the time, but Jim still found that although he was able to save some money, he would never get rich. Also, he began to be bored by the monotony of living in the isolated mountain community. He did get down to Sacramento once a year, but that entailed a long stage ride to Auburn and a journey by train; or a stage to Marysville, then by river boat down the Feather and the Sacramento Rivers. The bright lights of Sacramento only increased his determination to get the money he would need if he was ever to go on to San Francisco to enjoy the excitement of a real city.

As he handled the sponge gold and the bars of bullion, Jim came to accept the more or less prevalent belief that "God did not put all that wealth in the gravel banks for the sole purpose of making a few rich stock holders back east a lot richer." He was, perhaps, not too bright, but a little mental arithmetic convinced him that his five dollars per day was too small a share of the thousands in gold the monitor he operated was extracting from the blue gravel channel.

At first, the idea of increasing his economic well-being by doing what he knew to be a common practice in the area jarred his mid-western conscience; but as time passed, he saw that his chances of being caught were not too great and that even if that did happen, he stood a very good chance of going free on a technicality. The worst that was likely to happen would be that he would lose one of the best jobs in the diggings; also, he would be black-listed and no other mine would hire him. He was also smart enough to know that the slickings were wrecking the valley below and that it was only a matter of time until hydraulicing was outlawed. Then his mining days would be over. Even now the mining companies were driving their crews to their limits to get every yard of gravel blasted off the cliffs and through the sluice boxes before they were forced to turn off the water.

Jim saw that another job as good as his would not come along again in his lifetime. He would be forced to breathe rock dust and powder smoke in the shafts and drifts of the hard rock mines. So like so many others, he decided to get his "stake" while he could.

His plan of operation was simple. He studied the habits of the guards who patrolled the boxes. Being on the clean-up crew, he knew when the monitors would be turned off to allow the crews to scoop the values from back of the riffles. Near midnight before the valves were turned and the roar of the monitors was silenced, he slipped down stream to a section where he knew the amalgam would be heavy. Then, after he had made sure that the guard was following his usual routine, he reached down into the box and scooped up a double handfull of sand and amalgam along with the freshly trapped nuggets. Then in spite of the abrasion on his arms and wrists, he repeated the process until the small canvas bag he carried contained several pounds of concentrates which he knew carried at least fifty percent gold.

The ordinary highgrader would have "burned off" the mercury over the fire in his cabin.[1] He would then sell some of the sponge gold to the nearest saloon for about half of its real worth, and proceed to drink it up. He would continue to bring in more gold until his drinking money ran out, then he would repeat the process once again.

Jim did not make their mistakes. He knew the danger of distilling mercury over an open fire. Besides, if he did much of this, some of the other workers were sure to find out what he was doing. If that happened, he knew that he would be expected to settle their bar bills from time to time or his days at the mine would be numbered. So he buried his treasure and added to it the night before each succeeding clean-up.

When the robberies of the boxes got so heavy

---

[1]In doing so, he would have run the real danger of inhaling some of the fumes, and developing mercury poisoning.

that the mine put very heavy wire screen over the sluice boxes, it only made Jim's pickings better. Now he had very little competition. The other highgraders looked for and found unprotected boxes. He had access to the keys to the locks which held the wire; therefore, it was a simple matter to make an impression and have duplicate keys made.

Came the day when the monitors were silenced forever, and Jim was out of a job. He was in no hurry to start enjoying his fortune. It might not look good to begin spending too soon. Besides, he was not too sure that he was not being watched by one or more of the company agents he knew had been employed for just that purpose.

Much against the advice of his friends, and to the amusement of the local "experts," Jim bought the old Bill Higgins Blue Channel Mine up near Washington on the south wall of the South Yuba River canyon. Of course the Blue Channel part of the name was entirely due to the imagination of Bill Higgins. He had heard of the rich panning on the slope below his mine in the early fifties; so he figured that there must have been a Blue Channel somewhere on that canyon wall. He started prospecting on the slope, and finding some indications of ancient gravel he began tunneling into the hillside, convinced that within a few feet he would find rich gravel just as had so many others under similar circumstances.

Year after year Bill dug away. He did hit a small pocket of gravel just often enough to keep up his enthusiasm, to keep the storekeepers paid for some of their groceries and blasting powder, and to talk them into giving him another grub stake when things got really tough.

Jim paid five hundred dollars for the claim. That was about the amount of channel gold Higgins had taken out during the past three years. As a mine, it was a very poor investment, but for Jim the payment was only the premium on the "insurance policy" he had just acquired. A policy which would give him a safe annuity of ten thousand dollars a year for the next five years. Anything he could take out of the mine would be only a welcome bonus, totally unexpected, but appreciated if it came along.

As the old adage goes, "Them that has, gets." Jim had hardly drilled, blasted, and mucked out a dozen rounds, when he found where a sudden turn in the ancient channel had caused the pre-historic river to lay down a gravel bar rich in gold.

Now Jim had another problem. The nuggets and coarse gold he was shipping to the Mint made the sponge gold he included in each shipment look out of place. After 1933, had he tried the same scheme to get cash out of his stolen gold, he would have been investigated by the Treasury Department immediately. Not necessarily because it might have been stolen gold, but because it must have come from gold held out when all gold in private hands was ordered sold to the U.S. Treasury by the Gold Act of that year. Jim, however, faced no such problem. He was operating a mine which was demonstratively rich in gold. Just how rich, and how fine the gold, was no business of the bureaucrats, and he would have promptly and properly thrown them off the claim if one should have had the nerve to show up for an investigation. His only real worry was that some local character might try to get a reward for the recovery of some of the gold he had pilfered from the Eagle Mining Company. He need not have worried, for while some of the locals probably had doubts about all the gold supposed to be coming out of old Bill's mine, no real evidence was available. Nothing was ever said.

Jim hired a dozen Cousin Jacks from the Grass Valley hard rock mines and pushed a tunnel into the blue gravel channel. He knew that he would have to watch the clean-up operations very carefully, and make sure that the locks on the screens over the sluice boxes were changed frequently. He, of all people, should know all the tricks.

On the other hand he knew that he could not stop a miner from slipping a nugget into his pocket if one should show up during the mining and the mucking. His philosophy was, "let them have a little; God knows I am paying them little enough for the work they do, but a ten dollar nugget once a week or so will keep them working hard and on the job rather than running all over the hills looking for a mine that will pay them ten cents an hour more, or one that has highgrade rock showing enough free gold to make life more interesting."

Jim operated the mine for three years. He took out a half million dollars in dust and nuggets. The sponge gold from the Eagle Mine paid all the expenses at the Higgins Mine. The half million was clear profit.

With that kind of money, Jim became a regular habitue of Montgomery Street in San Francisco. He was wined and dined at the Palace Hotel and at the other first-class hostelries by innumerable stock and bond salesmen. He invested his money in what appeared to be gilt-edged securities, built a fifty thousand dollar house on Nob Hill, married the daughter of another successful miner and stock speculator, then settled down as one of the respected financial tycoons of the West.

Life seemed secure for the man from the streets of Chicago. His decision to chase the elusive gold of the Sierra had been successful [although not in the manner he had dreamed of in his fantasies] but Fate has mysterious ways of intervening in the course of human events.

A sharp financial panic of the late eighties turned the investments Jim had been assured were as good as government bonds, into paper as worthless as the Insul stocks of the 1930's. The main ledge of his father-in-law's mine was suddenly faulted and lost. His securities also became worthless. He was of no help to Jim.

The fair-weather friends from Montgomery Street were good for one last round of drinks at the Palace Bar, and "that was that." "The wheels of the gods grind slowly, but they grind exceedingly fine!"

Jim finished out the remainder of his life as he had begun, driving a street car; only this time he was the "grip man" on one of the cable cars which climbed the cliff-like hills of San Francisco, rather than whipping up the motive power of the Chicago horse cars.

Jim was philosophical about the whole thing. He had tasted life to the full. He had escaped from the drab flat lands of the Midwest to the romantic fog-bound hills above the blue waters of the Bay named for the good Saint from Assisi. On the days when the fog would allow, he could look far out beyond the Golden Gate and dream of the adventures which awaited there. "But, what the Hell? He associated every day with a group of the most interesting riders in the whole of this world. Who could ask for more?"

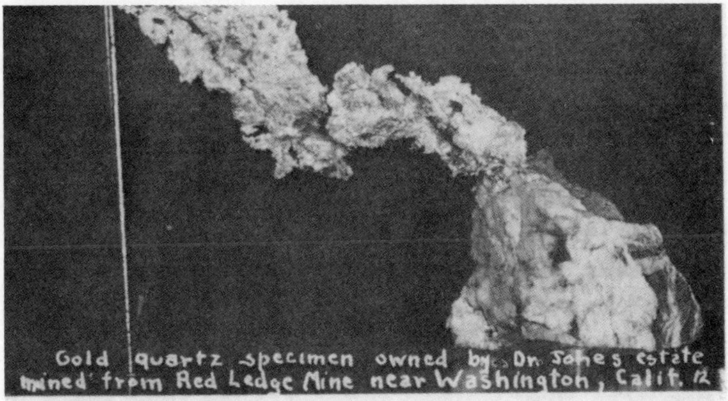

Leaf Gold attached to Quartz — Dr. Carl Jones Collection

CHAPTER XII

THE DREDGERS

At the time that the State and the Federal Government put teeth in the Anti-debris Law, forcing the closing of the hydraulic mines, the great dredgers took over to continue the flow of gold, keeping the economy of the country booming.

The modern version of the dredge was perfected in Australia. Before the floating dredger concept developed "down under," various attempts had been made to get the fine gold known to be in the gravel in the channels and on the plains where the Feather, the Yuba, and the American Rivers broke out of the foothills into the Sacramento Valley. There thousands of acres of gold bearing gravel was available for the taking. It contained a few cents to as much as a dollar a yard of the fine gold which had not been trapped in the bars and benches of the river canyons above.

These deposits had not been, and could not be mined by hydraulicing, because there was no place to take and store the cubic miles of tailings which must result. In addition, the land was too flat to provide the grade necessary to carry the gravel through the sluice boxes for the separation of the gold. The mountains where the hydraulic mines operated so successfully provided that grade.

Many methods had been used over the years to handle the gravel cheaply and to dispose of the waste material. The problem had been made more difficult by the hydraulic mines which operated in the watersheds of the rivers above. Often as much as ninety feet of slickings from the mines were piled on top of the "pay dirt" on the valley floor.

The great sheet of sand and gravel swirling out of the Sierra streams covered the mines, buried fields and orchards, and had an even more destructive effect on the navigable rivers. No shipping could go above the tide water at Sacramento. The filling of the channels stopped all but the smallest boats from sailing upstream except during times of major floods; and even then, navigation was almost impossible because the rivers were spread out over miles of the valley floor, making the main channel most difficult to follow. The government was forced to spend, and continue to spend, millions of dollars each year in its attempt to contain the rivers behind dikes, and to keep the channels open by dredging the mud from the bottom and piling it along the sides. The resulting dikes grew higher and higher until they could no longer withstand the hydrostatic pressure of the water on the river side. They broke, and the valley floor was flooded once again. In 1885, the City of Sacramento gave up trusting the high, man-made banks to protect the downtown section from the rising floods. It hauled in dirt by team and wagon sufficient to raise the level of a square mile of the "old city," by fifteen feet. All the streets were built up and repaved. All buildings were either razed and rebuilt, or they were jacked up and new foundations built on the filled land.

The first attempt to work the low grade gravels of the out-wash plains came soon after steam power was made available in the gold fields. Clamshell buckets were used to scoop up the gravel, then drop it into a hopper from where it was fed into a screening trommel to eliminate the rocks and small stones. The sand was then run through an ordinary sluice box to recover the gold.

This and similar devices worked very well on a very small scale, but it was not profitable to work the low grade gravel because an additional step had to be taken. The great volume of tailings (washed gravel) had to be taken away, which necessitated additional handling and hauling. In addition, the whole set-up had to be picked up and moved as the range of the boom handling the clamshell bucket was reached. This was always a time consuming and expensive procedure.

The miners in Australia had a different problem. Their gravel deposits were almost always found in the desert regions. Water was their major problem. To solve this, they hit upon the idea of putting the trommel on a barge which they then floated on a pond made by putting water in the hole they had already dug in their attempt to get down to the gold on the bedrock. Now the derrick with the "bucket" sat on the bank above the washing machinery. It scooped the gravel up stream from the barge. It was picked up and moved periodically as the gravel within its reach was washed. The barge followed. But, because it floated, it was cheap and easy to move.

The problem of the dredger operator was still not completely solved. The tailings still had to be disposed of. At each new set of the barge, the tailings presented no problem. But that lasted only a short time. The pond area behind the barge soon became filled. A second derrick and bucket was added to lift the tailings up and place them on piles. Unfortunately, the piles soon became so high that the larger rocks spilled down the sides and threatened to swamp the barge.

In spite of its faults, the Australian dredger was imported into California and put to work washing the richer gravel along the flood plains of the rivers in the Central Valley.

Almost immediately the Australian dredger was

improved. The derrick and the bucket at the head of the barge was eliminated. Instead, an endless chain of buckets was lowered from the upstream edge of the barge. A steam engine supplied the power to lower the buckets to the bottom of the gravel, scoop up a load, elevate it to the hopper above the screening trommel, dump its load and return for yet another load. The whole process was repeated continuously. The barge or boat was anchored to the bank by cables which could be slacked off or tightened by capstans.

As so often happens, one improvement in a mechanical process makes for yet another problem. In this case, the great efficiency of the endless chain of buckets which delivered the gravel to the washing machinery also produced such a great volume of tailings that it became impossible to handle them with derrick and bucket.

But all problems have solutions. Some just take longer than others. In this case, the gravel, minus the gold, was spilled onto an endless leather (later rubber-coated canvas) belt. It traveled up an inclined plane to the desired distance. At that point the gravel spilled off the belt to form great conical piles.

Again, to keep the piles from becoming so high that the larger stones would roll down the slopes and threaten to sink the boat, still another refinement was added. Now the boom, which carried the tailing belt up the inclined plane, was itself hinged to the downstream end of the boat. This allowed the boom to swing from side to side in a semi-circular arc. The boat was advanced in successive steps, not continuously. This led to the quarter circle piles of tailings which came to cover thousands of acres of rich valley farm lands which had been some of the most productive in the world—as rich as those of the Nile Delta!

As long as the dredgers operated on land

already devastated by the tailings from the hydraulic mines, no great harm was done. Once this land had been "eaten," and the resulting "droppings" left to scar the landscape, and the channels of the "living" rivers washed free of the gold the sluice box operators could not get, the land-eating monsters looked for more land to devour. It was found next to the land that had already been turned into great unsightly windrows of river rock.

The dredging companies purchased the land on either side of the land they had already devastated and proceeded to render useless for all time (supposedly), highly productive farm lands.[1]

A short drive out of Sacramento, or out of Oroville, takes the observer through a mini canyon cut through the tailing fields, fields which once were green with alfalfa or set with rows of almond and peach trees. Now the only vegetation to be seen is an occasional clump of hardy grass or an even more drought-resistant, isolated creosote

---

[1]The thousands of acres of tailings which once carpeted the area between Sacramento and Folsom have just recently become valuable once again. The population explosion in the Sacramento area has placed a premium on living space. Land developers have found it profitable to bring in giant bulldozers. The semi-circular windrows of tailings have been flattened and six inches of top soil spread over the pavement of cobble stones; now, shopping centers, apartment complexes, and thousands of single-family homes are built on the land which until ten years before was "known" to be of absolutely no use forever! (One can only pity the innocent buyers of this property, especially those from the eastern states, when they begin to plant their gardens only to find that under the thin layer of soil the subsoil is but a pavement of cobble stones.)

bush, each having found a precarious foothold on a bit of surface soil which had, by a miracle of chance, escaped being buried under fifty feet of coarse gravel and boulders.

During the hundred years the acres of tailings baked under the hot California sun, some slight uses were found for the billions of tons of the river bed rocks. The first and the most obvious was for crushed rock for the construction of foundations and for the pavement of our highways. Later, when the great earth-fill Oroville Dam on the Feather River was built, and the Folsom Dam was thrown across the American River, eighty percent of the fill consisted of the materials spewed out by the gold dredgers. At Oroville, in particular, a railroad was constructed from the dredger fields below the city to the dam site five miles up the canyon. Special cars and special loading machines were built to allow the trains to roll by the loading facilities without stopping, then continue on to the dam site. The amount of material required to fill the space that is now the dam was so great that the trains operated 24 hours per day over a period of nearly four years before the dam was completed; and yet the materials available were so great that one can hardly notice the loss of the rocks from the piles left as the dredgers moved on.

The Folsom Dam used the tailings from the American River below Folsom. But the dam was so much smaller and the amount of the tailings was so much greater, that there is no evidence that any were used.

An unexpected use of the windrows of tailings came in the early 1950's. The Rocket Engine Division of the Aero-Jet Corporation was forced to leave Orange County in southern California. The blasts of the test runs of their rocket jets became so much of a nuisance that the local residents "ganged up." The County Zoning

Commission, knowing where the votes were, ruled against the nerve-wracking tests.

The Aero-Jet people looked for a test site where they could purchase a large tract of land immune to subdivision sprawl, yet close enough to a metropolitan area so that their thousands of employees could be housed. The great dredger tailing fields south of the American River seemed to be the answer. They purchased several thousand acres of the waste land for little or nothing (probably for back taxes only).[2] Bulldozers quickly leveled off the sites for their engine building plants, and roads were cut through the rock windrows to the test sites. An unexpected bonus came when it was discovered that the curved piles of rocks fifty or more feet high, acted as natural baffles to absorb the shock waves from the engines being tested. The driver along Highway 50, or Interstate 80, could often see great mushroom clouds of white smoke rising from the rockets being tested, but he would not hear sound, nor would he be buffeted by the shock waves which would have developed on normal flat lands.

As with the hydraulic mines, the conservationists attacked the dredgers for the wanton destruction of our most precious national resource; our rich farm land. Efforts were made

---

[2]The population explosion, in part, at least, due to the industrial development led by the Aero-Jet Corporation, leveled the natural baffles, and once again the jet engine testing had to be moved. But as usual, "the rich get richer." The land which the company had thought good for nothing but to add distance between their operations and "civilization" and had cost it next to nothing, suddenly became prime subdivision land. (Some of it after being leveled has recently sold for as much as $150,000 per acre! The company could well afford to move their test site elsewhere.)

to either stop the dredgers from ruining any more good crop land, or to work out some means of saving the top soil and putting it back on the surface once the tailing gravel had been deposited and leveled. Laws were passed to force the dredger operators to do this, but they were never enforced. The various counties put what they considered prohibitive severance taxes on the land if it was to be destroyed, but the investment in the dredges and in the land, as well as the fact that the counties had underestimated the profits made by the dredging companies, had no effect. They paid the tax and continued to operate.

Only the bombing of Pearl Harbor stopped the dredgers. The government ruled that theirs was a non-essential industry. They were forced to stop the endless chains of buckets from eating yet more land. However, this was only temporary. Once men and materials became available again, the dredgers began destroying the land once more. In fact several new dredgers were built and put to digging for more gold along the flat lands between Marysville and Oroville.

At last the "laws of economics" did what the politicians could not do. The rise in the cost of supplies, the demand of the men for higher wages, and the fixed price of gold caused the balance sheets of the companies to end in the red.

All but a few of the dredgers stopped operations. They were not allowed to lie idle and deteriorate; instead, they were dismantled and shipped to the raw undeveloped valleys of Canada and Alaska, to the lush river bottoms near the head of the Amazon River, and to the previously unexplored interior of New Guinea. In almost every case, crews were sent ahead to brush out and to build a landing field. The dismantled dredgers were shipped to the nearest sea port with a nearby air field, loaded aboard a freight plane and flown to the newly constructed air strip, unloaded, reassembled, and put to work in a pond dug by a previously flown-in bulldozer.

In the Arctic, the dredgers have an additional problem. There the permafrost makes the gravel almost as hard as solid rock. A steam plant must be added and the ice melted before the gold bearing gravel is scooped up by the buckets.

In the tropics, the main problems are the mosquitoes and the blow guns of the head hunters.

The highgraders were pretty well "stymied" by the dredgers. The sophisticated, vibrating sluice boxes were covered by steel gratings which, in turn, were covered by the superstructure of the barge and were accessable to the crew only. The only time that the riches were exposed to anyone with sticky fingers was during the clean-up period. The great machines were operated on a 24-hour basis, stopping only for repairs to be made. During that time, the concentrates were taken from the recovery machinery. The gold was melted and cast into fifty pound bars and shipped to the Mint.

Bars of bullion were almost impossible to dispose of. In a few cases holdups were staged, usually when the gold was on its way from the dredger to the post office. Most robbers, however, were not familiar with raw gold. It proved to be much easier and probably safer to hold up a bank and come away with spendable cash.

The only times that gold in any appreciable amounts was highgraded from the dredger operations was when the superintendent of the clean-up crew got the idea of sharing the wealth. Any smart clean-up boss could hold back a few lumps of sponge gold before it was dropped into the crucible for melting. A little melted gold could be left in the bottom of the crucible, and of course, it was not always possible to pour the melt into the molds without spilling a little. True, all sweepings were carefully smelted to recover any overflow or spilled gold, but even just a few handfuls of sweepings shoved into a pocket rather than being dropped back into the retort could amount to a good many ounces during a year's time.

The dredger operators watched for this simple type of highgrading. The usual procedure was for the company to hire a "private eye" and give him a job on the dredger. Of course, all "extra-curricular" activities ceased when any new man came on the job. Once the newcomer had been checked out, the highgrading began once more. If it was suspected that the new man was a spy, his movements were carefully charted, and ways to get at the gold when he was not around were worked out.

Duplicate keys to the box covers were made. Holes were drilled in the bottoms of the boxes and plugged until an opportune time arrived to pull the plugs and catch the amalgam as it dropped through.

There were always ways to outwit the company's men. Sometimes it worked both ways. The watcher drew his regular wage and his bonus for watching the others; then after a suitable time began working with them, taking his share of what the gang took from the clean-up operation.

The companies knew that this was apt to happen. They used other methods. Instead of putting their man to work, they paid him to sit at the bars of close-by roadhouses such as the Seven Mile House east of Marysville, and the Folsom Nugget in the town of Folsom. He was paid to observe the spending habits of the dredger workers, and to listen to all the local gossip. He knew the honest income of the men he watched, and any unusual bar bills were reported to the company at once.

Unions were no problem. Organizers were just not allowed near the dredger fields; so there was never any question of job tenure. If Bill Jones seemed to be spending more than his daily wage would warrant, he was simply dismissed. If the private eye was absolutely sure of his evidence, Bill would be black listed and immediately become no longer eligible for dredger work. Injustices were often the result of such ruthless management methods. The tenor of the times, however, were such that only a few enlightened men ever gave it any more than a passing thought. Perhaps such utter disregard for the workman's rights was one of the reasons the present Miners, Mill, and Smelter Workers' Union was, until very recently, at least, considered to be the most radical labor union in the United States.

With the advent of the automobile the workers were scattered. No longer were they forced to live within walking distance of the dredger. This complicated the work of the investigator. It also increased the temptation of the worker to add a little to his daily wage. Certainly, on the wages paid, the ordinary worker could not buy and operate a high powered automobile. The superintendent and the engineers might be able to afford to purchase a new car as the new models came out, but the workers had to be content with a second hand "klunker." The purchase of anything other than a Model T Ford off the show room floor was an almost sure way for the worker on a dredger to get fired.

After 1933, another obstacle appeared to check the operations of the highgraders. The Federal Government entered the picture, and the source of all gold sent to the Mint had to be sworn to. Now the highgraders, not only had to outwit the guards and the detectives of the dredger companies, they also had to find a buyer for the gold. Many and varied were the schemes worked out to get the gold to the Mint without arousing suspicion. Because these methods were basically the same, some of the subterfuges will be explored along with the stories of the highgraders who were disposing of the quartz mine gold.

Serpentine Rock & Digger Pine Country

Finnish crew at the Mattie Mine about 1933 — Mike McGuire standing

CHAPTER XIII

THE DOODLEBUGGERS

Is there any future for the placer mines of California? The fact that there is more than $30,000,000,000 (at 1979 prices) locked in the gravel banks and in lake bottom deposits would seem, to the uninformed, cause for great optimism; but locked is the proper term. Only a tiny fraction of the great wealth—wealth which is in excess of the value of all the gold mined in the Golden State from the time of Marshall to the present—can be recovered by any method other than by hydraulicing. Barring some natural disaster which would warrant the filling of all the reservoirs on the lower reaches of the Sierra rivers with hydraulic slickings, there is not the slightest possibility of that method of disposal of the wastes from the mines being allowed. Certainly the economists who believe that gold has outlived its usefulness as a basis for money, would never advise the politicians that it would be beneficial to wash out the gold from the California mountains only to rebury it at Fort Knox in Kentucky.

A major depression would send the "snipers" back to the river bars just as it did in the 1930's. At that time twenty-five to thirty men could always be seen working their Long Toms and rockers on the river gravel where the Yuba River bridge crosses that stream as it emerges from the foothills into the flat lands of the valley. A thousand more were scattered along the upper reaches of the river searching for the gold washed down by the sixty floods of the winters since the closing of the hydraulic mines. The more optimistic were also searching for the gold the '49ers, and the Chinese who had followed them, had missed. Every experienced miner knew that not much was missed by the men who had been there before, but as the poet is so wont to say, "hope springs eternal in the human breast."

The pessimists were for the most part correct. The snipers of the thirties had to be satisfied with wages that even the coolies of the 1870's would have despised. Most of them made only fifty cents to one dollar a day, and the price paid for the gold they mined was double what the Chinese received.

A surprising number of the Depression miners were college graduates. They lived in their "Hooverville shanties," ate beans and sowbelly, and went through most of the hardships of the '49ers. They suffered, not because they expected to strike it rich, as did their predecessors, but because that was their way of showing their independence and their determination to stay off the relief rolls. Unless they were actually starving, they were not going to "sponge" off the tax payers. They were the true sons of the Argonauts.

Many people, looking at their beards (this was before the days of the beatniks and the flower children of the sixties), their ragged jeans, and their willingness to live in brush and cardboard shacks, were inclined to think that they had lost all their pride and would never be of any value to society again. What their critics did not know was that these men were fighting back in the only method they knew. They were keeping in fighting trim, both mentally and physically, so that when the opportunity presented itself, they would be ready to contribute once more.

W.C. Way (no one ever did find out what the "W.C." stood for) was a typical example of the men with this determination. He walked into the laboratory of the Chemistry and Physics Department of the Yreka High School one day in the fall of 1933. He was poorly dressed, but his clothing was clean. His white beard was neatly trimmed, and it was not surprising that he spoke with a slight Oxford accent.

Exactly what he had come in to inquire about has been forgotten, probably something about how much cooling surface he would need if he built his own retort for the recovery of the mercury from his amalgam. Later he came in to see if there was any known reason why the extremely fine gold seemed to adhere so tightly to the corrugated rubber mats from off the floors of the junked cars at the Yreka city dump? He had salvaged several of these and was using them to catch the very light flaky gold in the sluice box he had built. He suspected that there was an electrostatic attraction between the rubber and the metal. He was looking for literature on the subject.

His Oxford accent was real. He was born in Australia; had a degree in electrical engineering from the University of London; and for many years had been chief engineer for a large public utility in the Pacific north-west. He had raised a family and had retired to open his own consulting engineering firm, secure in the knowledge that his Insul Utility stock would allow him to live in comfort whether or not anyone ever asked for his professional services. That was in 1928. By mid-1930, he was dead broke. He had the clothing on his back, a few tools, some camping equipment, and a 1922 Dodge panel truck.

Way could have had a supervisorial position with the WPA, bossing a leaf raking job, but he was made of sterner stuff. For gas and food, he sold his last few pieces of furniture for ten percent of what he had paid for them only two years before, and set out for the mining country along the Klamath River in northern California.

He was doing much better than most. He had found a small bar on Humbug Creek a half mile up from its junction with the Klamath. There, the '49ers had taken out a minimum of a dollar a pan, and ten dollars was not uncommon. Way worked ten hours a day, six days a week for less than a half ounce of gold in the form of very fine dust.

On Sunday, like the '49ers, he did not work, but cleaned up his camp, washed his clothes, and cooked up a pot of beans for the coming week. In the little time remaining he tramped the hills, in part for recreation, and partly to see if he could be lucky enough to stumble on something the earlier miners had missed. He did find something. Not much, but it was better than the gravel he had been washing.

About two miles on the hillside back of Hawkinsville, he found a quarter inch stringer which had pin-head-sized chunks of free gold scattered throughout. It took him almost a day to pick out a panful of the sugary quartz. Then he pounded it in a four quart steel mortar. This done, he panned it to recover the gold.

There was nothing he could do about the mining operation; it was just hard work with a pick and shovel. The quartz stringer was much too small to drill and blast. All the quartz would have been lost in the rubble brought down. Fortunately, the country rock was relatively soft so that it could be removed with a sharp pick wielded by a determined man; but there was something that he could do about the crushing and panning. He used some of the same kind of ingenuity used by the '49ers in developing the early mining machinery.

Way became a regular visitor at the Yreka wrecked auto dump. He salvaged some heavy channel irons from an old truck. These he used as a base. Five truck axles with their original gears filed and flattened by an arc welder took the place of stamps. A tireless rear wheel of a 1924 Dodge with the flange in place acted as a pulley, and an additional axle with eccentrics welded at the proper places acted as a camshaft to raise and let fall the make-shift stamps. Wooden guides held the stamps in place. A length of garden hose brought the needed water to the base of the stamps, and a screen held back the material not ground fine enough to release the gold.

He bolted a cast off pulley from a 1910 steam tractor engine to the rear wheel of his truck, jacked it up, ran a belt around the truck pulley and over the stamp mill pulley, and he was in business. Of course, he had to feed the mill by hand, but the grinding problem had been solved. For a total of five dollars at the local blacksmith shop, and a lot of ingenuity, he had put himself in a position where he was the envy of every miner up and down the Klamath.

The gold he was able to recover by running the mixture of water and ground up rock over a slanting table on which he had tacked several lengths of corrugated rubber mats from the floor boards of wrecked cars. Later he was able to buy a four by five foot sheet of copper. This he coated with mercury and placed below the rubber matting to catch any gold that had not been trapped by the corrugations above.

After Way had worked out the stringer on Hawkinsville Mountain, some of his friends grub-staked him so that he could open up an abandoned mine in Quartz Valley, some ten miles northwest of Fort Jones on the Scott River. The ledge in that mine also pinched out after a time; so in 1938, he got a job testing gravel for a dredging company. That job lasted until the defense industries opened up in the San Francisco Bay area. At that point, he had no trouble becoming the chief electrical engineer for one of the shipyards.

Perhaps Way would have been just as well off

if he had lived on a relief check during the eight long years he had grubbed in the quartz mines and in the gravel, fighting to stay alive during the Depression years. But it is much more likely that eight years of idleness would have been most devastating for a man of his engineering backgound. He might not have had the ability to become a man of importance once highly skilled men were again in demand.

Not many of the Depression Argonauts were as successful as Mr. Way. But they did have one thing in common, their pride. They went to every extreme to keep off the relief rolls and to maintain their self confidence.

While Way was developing his home-made stamp mill out of discarded auto parts in Siskiyou County, hundreds of others were working to improve their primitive methods of working the gravel deposits up and down the length of California. With a little money, some imagination, and security in the knowledge that a lot of hard work never hurt any man, the most ingenious devices were put together. Some were labeled "Rube Goldbergs." But the name that eventually stuck was "Doodlebug." The operator of such an outfit was a "Doodlebugger"; or if he was operating on, or prospecting various gravel deposits to see if there was enough gold to make it profitable to move in heavier machinery, he was known to be "Doodlebugging."

The Forest Service had opened up much of the back country with fire roads, and the CCC boys had made more of the mining country available by building more fire and logging roads. These could now be used for trucking heavy equipment into heretofore inaccessible canyons. A man with a little money, or was more often the case, a group of men with a few dollars each joined together, could buy a dump truck, a bulldozer, and a rotary gravel screen from a bankrupt road builder;

overhauled, or rebuilt the machines, and the miners were in business.

With a little luck it was possible to find an isolated bench of gravel that was too low grade, and too far from water for even the Chinese to have worked. The bulldozer of the Doodlebugger cut a road from the gravel to water. The machine was then used to push the pay dirt into a loading chute. The dump truck hauled the gravel to the screen, now called a trommel, where it was washed, and the fine material sent through a sluice box for the recovery of the gold.

The advantage of the Doodlebug was that it was very mobile. The same equipment which was used to operate it carried it from place to place, making its own road where necessary. A natural variation was to set the trommel on a gravel bar of a river and use a power shovel or a loader in place of the bulldozer blade to put the gravel into the revolving screen.

Using that method of Doodlebugging, it was possible to work gravel with a very low gold content. Gravel paying as little as ten or fifteen cents per yard proved to be quite profitable, and twenty-five cents a cubic yard was a real bonanza. Not too much of this was available because, with that much to be extracted, it could be worked by hand with a Long Tom. The earlier miners took that gold. The Depression miners had to be content with gravel of a lower grade.

After the hydraulicing was stopped in the 1880's, the gasoline engine was developed, along with high pressure centrifical pumps. It was natural, then, that someone would put the two together and start blasting at the gravel banks again with powerful water jets. Of course, they could not foul the streams with their waste rock, so they used their bulldozers and their dump trucks to build check dams below their workings.

These dams kept the debris from the stream channels; and, at the same time, impounded the water so that it could be used over and over.

This was exactly the method used in blasting away the point of the hill and building up the floor of Strawberry Canyon for the construction of the Memorial Football Stadium just east of the University of California in Berkeley. It is said that "necessity is the mother of invention." Was it invention, or was it just some depression-hit, ex-student from Berkeley adapting the earth moving idea seen in his student days to the necessity of the moment?

Few had the capital for a successful Doodlebug operation. Mostly the depression miners were reduced to their last few dollars before trekking to the mining country. Lacking capital, they were forced to depend on the pan, the rocker, and the Long Tom to get the meager amount of gold they needed to stay alive.

The gold they could recover was extremely fine, usually so fine that it had floated away from earlier operations. The Argonauts of the 1930's depended on being more skillful and more careful than were their predecessors. Consequently, they built their sluice box riffles with extreme care, and they adjusted the slope at which the box was set, and observed with more care the amount of water going through the boxes to see that it was just enough to carry the sand through, but at the same time give all the gold time to settle behind the riffle cleats.

A woolen blanket had been one of the first refinements used by the early stamp mill operators to recover the very fine gold. After the sand and the water had passed over the riffles or over the copper plates, careful panning still showed a small amount of flour gold being lost. One of the earliest of the California miners must have been a student of the classics. He had read of Jason and the Golden Fleece (probably in the original Greek). He knew that the Greek hero was attempting to get to the rushing rivers as they tumbled down the slopes of Mount Ararat. There he knew that the native miners were suspending sheep skins with the wool still attached in the swirling flood waters carrying flour gold down off the sides of the mountain. The gold was trapped in the kinky, wooly hairs. The skin was then removed from the stream and the gold rinsed out in a container of clear water—hence the "Golden Fleece."

The miners of the thirties reintroduced the idea of the loosely woven woolen blanket to catch the fine gold that had escaped all other methods of catching it. Then they added an invention of their own. They found that a corduroy coat could be spread out below a rocker or a stamp mill and the very fine gold trapped by the ridges of cotton fibers. This proved to be even more efficient than the woolen blanket.

The '49ers did not have corduroy to use, and they had another disadvantage over the miners of the thirties. Their only method of recovering the gold after it had been trapped was to rinse the cloth in a tub of clear water, then pan the settlings.

The Depression miners had a better way. They allowed the blanket or the corduroy to dry then they burned it and panned the ashes. The '49ers could not do that, as the cost of the cloth was often more than the value of the gold recovered. During the Depression, the miners had no such worry. A quick trip to the dump near any city or town provided all the worn blankets and corduroy pants and coats they could possibly use.

Again the burning process was not new. The hydraulic miners used it. Because of the heavy abrasion of the wooden materials in the sluice boxes, it was necessary to constantly rebuild them. The old lumber was saved, carefully dried and burned; then the ashes were panned. It was surprising how much fine gold and amalgam worked into the tiny cracks where the sides of the boxes were nailed to the bottoms, and into the openings where the cleats had pulled away from the planks below.

Since the Depression there has been only one innovation in placer mining. The snorkel and scuba divers have learned that they can have their fun, and in a few cases have a well paid vacation. Both work under the surface of the water directing a vacuum hose from a pump mounted on the surface, along the bottom of the Sierra streams. The pump draws up sand and gravel along with the water. The pumped material is delivered ashore into a sluice box. Any gold found mixed with the gravel is recovered.

The early miners, both the Americans and the Chinese, were very adept at getting at the gravel along the river channels. They did it mostly by diverting the water from one side of the canyon floor to the other. The dry channel was then scraped free of its gravel and the river returned to its original bed. In a few places the solid rock walls on either side of the running water did not allow a new channel to be dug. The holes there have never been prospected. The modern scuba divers search for these spots along the rivers. They are equipped to get at the gravel on the floor of the deepest holes.

Often the scuba miners are disappointed. They find a deep hole which would seem to be so isolated and so enclosed by rock walls that no one could have diverted the water away from it. Alas, they find the bottom swept clean of all but the small amount of gold which has accumulated since the time of the '49ers. Careful searching will usually show that they credited the earlier miners with too little ability. Usually it will be found that the diversion dam was built much farther upstream than expected, and the low water of early fall was carried in a ditch high up on the canyon wall, well above the rock outcropping responsible for the deep hole in the first place. A natural phenomenon may also cheat the scuba miner. The swift currents of the river or creek may have carved the deep hole out of solid rock leaving the floor of the hole smooth and unfractured. In this case, there are no obstructions to stop the gravel and its gold from being carried up over the lower edge of the pool. The diver finds only a smooth, well-swept floor.

The river and creek diverters were also apt to be misled into believing that there was wealth at the bottom of a hole which apparently had not been drained and swept.

In 1923, Mike McGuire and two of his friends spent nearly a whole year building a diversion dam a short distance above the falls on Deer Creek three miles west of Nevada City. They then dug a ditch around the head of the falls and back to the creek three hundred yards below the pool at the foot of the falls. With the main channel dry, they pumped the water from the pool expecting to find a supply of very rich gravel.

They had made one serious miscalculation. A very ancient flow of basaltic lava had ceased flowing at about this spot. The sudden ending of the very hard basalt had been the cause of the falls in the creek channel. The rock below was a more easily eroded granite. The pool that the falls had formed was in a non-fractured section of this granite. There were no obstructions to catch and to hold the gravel and hopefully the gold it carried.

Mike and his friends got no gravel to pan. Their sole payment for all their effort was a single silver coin, a quarter, which someone had recently lost in the creek.

Gold and the Highgraders

During the 1850's and the 1860's, the large mining companies spent millions diverting water away from the channels of the rivers. Many of these diversions were miles in length, and the amount of gold recovered was great.

One can only speculate as to the number of nuggets the hundreds of mine employees put in their pockets as they scraped the river beds where the gravel was richest. Probably not more than one percent of these '49ers could resist the temptation when a golden nugget rolled out and no one was looking. And who could condemn them? After the hardships they had undergone to reach the mine fields only to be disappointed in their own search for gold and then being forced to go to work for two or three dollars a day for a million-dollar corporation, they felt justified in taking a nugget or two. At the very least they had to get a few nuggets for themselves to show the folks back home that they actually had found gold.

Even the well-capitalized mining companies did not always make money by changing the course of a river. The gold bearing rivers of California are unique in the tremendous fluctuation of their flow from month to month. The river diverters began their work digging their ditches and building their flumes as early in the spring as weather permitted. As the summer progressed, their work was carried feverishly foreward. By the middle of August, the snow water from the high mountains had been carried to the valley below, and the flow of the river had dropped from a volume in May of perhaps 100,000 cubic feet per second (on Feather River for example) to a flow of less than 200 cubic feet per second. This relatively small flow was easily turned into the now completed ditches by a small amount of dirt and rocks thrown across the river channel.

The ditch diggers now turned miners. They swarmed into the river bed, pumped out the deep holes and shoveled the gravel and the scrapings from the cracks in the bedrock into the already set sluice boxes. They literally worked night and day getting out every bit of gravel before the fall rains began and floods swept down off the rocky ridges above to sweep away the diversion dams and to fill the dry channel once more.

An early rain in the high Sierra could ruin most of the costly summer's work before the gravel could be recovered and washed. The first high water took out almost all the ditches and flumes. If the channels were not cleaned before the first rush of water, most of the work would have to be redone the following year.

The risks were great, but so were the rewards. The gold receipts at the San Francisco Mint during the 1850's and 60's showed that the river miners were delivering between forty and forty-five million dollars of channel gold annually. It is of little wonder that the scuba miners find it increasingly difficult to locate unworked river beds.

So the last bit of placer mining, so suddenly revived by the scuba divers will soon have run its course. With that day will come the end of the placer mining in the Sierra. It will have arrived, that is, unless science finds a way of mining the thirty billion dollars of gold still locked in the great gravel banks left by the hydraulic miners when they were forced to close because of the damage they did to the valley below.

CHAPTER XIV

BECKWOURTH

At the time of the discovery of gold in California, the Sacramento Valley had the only beginning of civilization other than along the coast, and along El Camino Real (the Kings Highway) leading from San Diego to San Francisco. Sutter's Fort at Sacramento and a dozen or so ranch headquarters occupied by American pioneers such as Bidwell at Chico and Ide at Red Bluff, were the first to settle on the 20,000 square miles of flat lands along the Sacramento River. The even larger San Joaquin Valley was, as yet, mostly unexplored wasteland.

A few Mexicans had filtered north from the Los Angeles Basin and had great herds of horses grazing the hills around their headquarters at Tejon (later Fort Tejon) on what is now the "Ridge Route" between Bakersfield and the San Fernando Valley. The size of the herd can be pictured when it is remembered that in 1846, Jim Beckwourth stole more than 5,000 head in a single horse stealing raid.[1]

Beckwourth (sometimes spelled and pronounced Beckwith) was the son of a Missouri mulatto house maid. His father was a former major on

---

[1]The facts concerning this bit of horse stealing are open to question. But there is no doubt that Beckwourth did steal a large number of horses from the Mexican ranchers; but according to Jim's own story, they came from Orange County near Los Angeles. This was after General Castro refused to pay for his services during the rebellion against the Governor in Monterey.

Washington's staff during the Revolutionary War, and at the time of Jim's birth, a successful St. Louis business man. He was also the legal owner of Jim's mother. Although technically a slave himself, Jim, with the consent of his father, ran away from the blacksmith to whom he had been apprenticed (after throwing a ten-pound sledge hammer at his master's head), and made his way up the Missouri River with the first overland beaver trapping caravan commanded by General Ashley, the organizer and owner of the Rocky Mountain Fur Company. He trapped beaver successfully for two years, and then, in spite of his youth and his color, he was chosen to lead a trading mission to a Blackfoot village on the Madison River.

The chief of the village had a daughter who was too sharp of tongue and too headstrong to be of interest to the village braves. Jim was the answer to the chief's prayers. Jim could not refuse the offer of the chief to become his son-in-law. Without more formality than the payment of two blankets and a pound of gunpowder, Jim found himself a married man.

The honeymoon was not too happy. At the end of the second week of married life, Jim was looking for a way out. The chance came in a most unexpected way. A few of the Blackfoot braves came charging into the village with the scalps of three white trappers streaming from their lances. Jim forbade his wife's joining in the celebration and the scalp dance which followed. His willful wife disregarded his orders. She slipped away from his lodge and joined the dancers.

Jim was in a quandary. His trading mission, which consisted of himself and two white squaw men, was at best only tolerated by the Blackfeet. One false move might not only drive them out of the village, but it could cost them their scalps as well. He asked the two old-time trappers what he should do. Their reply was that they could not tell him just what to do; but he had damned well better do something, and quick, or he would be the laughing stock of the mountain men from then on.

Now Jim had not won his promotion to the head of the trading mission over the old-timers because of timidity or through lack of action when it was called for. He had repeatedly held his own in shooting incidents and in hand-to-hand combat with the mountain men and with the Indians. History repeatedly shows that his temper was set on a hair trigger. So he grabbed his errant wife by the hair, dragged her out of the circle of dancers, and bashed her behind an ear with the flat of his tomahawk. She went down like a steer in a slaughter house.

The whooping, dancing braves came to a sudden stop, as, for an instant they stared at Jim with open mouthed astonishment. For his part in the continuing drama, Jim turned his back on the crumpled body, and on the stunned savages, and with an air of contempt stalked back to his lodge.

The other traders joined Jim in his tepee and prepared to sell their own scalps as dearly as possible when the young braves started to carry out their threat of scalping the whites while they were still alive and then to finish off the job by burning them in the council fires. Chief Broken Arm, however, had other ideas. He turned his Dog Soldiers (village police) loose on the blood-thirsty warriors and held them at bay until he could show them the folly of their ways. He pointed out that his daughter had been a bad wife. She had disobeyed a direct order of her husband. She had gotten exactly what she had deserved. There was not a brave in the Blackfoot nation worthy of the name, who would not have done the same if any of his wives had disobeyed.

"Now," he said, "My heart grieves for my daughter, but more than that, my heart bleeds for my friend the white trader. My daughter was not a good wife. His lodge will now go unswept and his bed will be cold. I will now give my second daughter to my friend so that he may sleep in comfort tonight. The white trader got a poor bargain when he got my first daughter. He is not to pay for daughter number two."

Jim breathed a great sigh of relief, both for the security of his scalp, and for the way the wife problem had been resolved. The second daughter had all the attributes that the first had lacked. She was an uninhibited savage beauty, but she was well trained in the duties of an Indian wife. Jim had tried to get her before his first marriage, but the chief seeing a chance to get rid of a product which had no local demand, had insisted that Jim accept the "damaged" goods.

The way things turned out caused Jim to congratulate himself. After a suitable wedding feast, he retired to his lodge and to bed with the belle of the tribe.

His joy did not last too long. About midnight he was rudely awakened by a second female crawling into his bed of buffalo robes on the opposite side from his new wife and snuggling up to him with an all too familar set of motions.

It seemed that the blow of the tomahawk had not been fatal after all. Wife number one had merely been stunned. She had lain outside his lodge, where the braves had dragged her, awaiting burial. She had regained consciousness and was trying to win her way back into the favor of her lord and master by the same method used by every woman from Eve on down through the ages.

Jim stayed in the village only six weeks more. Wife number one was becoming troublesome once again, and the Blackfeet had found that their interests were best served by dealing with the Astor traders—the American Fur Company. They declared open season on all whites not associated with that company as well as on all their Indian neighbors of whatever tribe. From that point on, they were known as the S.O.B.'s of the mountains and of the plains; a condition which was to exist until the Sioux took over following the smallpox epidemic of 1837-38, which wiped out three-fourths of every tribe along the Missouri, and in the case of the Mandans, killed every man, woman and child in their main village.

Jim found that the Crow Indians were more to his liking. He again moved into an Indian village. With the Crows he did more than trade. He joined their hunting parties, their war parties, and cooperated with their horse stealing raids. Within months, he gained the reputation of being a great warrior and expert horse thief.

During the first brush between the hunting party he was accompanying and a Blackfoot party on the prowl for scalps, he nearly lost his life before his career had hardly begun. But by displaying a great amount of courage and leadership, he turned what appeared to be certain death into a victory.

The traditional method employed by the Indians was for the opposing sides to line up their horses in long straight opposing lines, then gallop head on, man against man, much as the knights of the middle ages rode each other down in their jousts. In the case of the Indians, the charging lines on both sides "chickened out" just before they got within effective arrow range. They then wheeled and tried to maneuver to encircle the enemy. That established, they rode in an endless circle with relative safety, while they released their arrows from under the necks of their galloping horses.

Jim must have been familiar with this method of warfare, but it was so foreign to his nature that on the first charge at the oncoming wall of Blackfeet, he refused to turn his horse at the prescribed limits. Instead he lashed his mount straight ahead, and rode headlong into what appeared to be certain death—one lone man against a hundred superb warriors, each armed to the teeth and screaming for blood. His audacity stunned the Blackfeet. They did not bunch up, but sat their horses in surprise. As Jim hit the line of warriors, his well trained horse swerved just enough to bump shoulders with the horse of the individual Blackfoot he had challenged directly. The contact threw both the horse and the rider off balance just enough to spoil the aim of the warrior's war club. Jim caught the blow on his buffalo hide shield, and crashed his own club into the Blackfoot's skull, scattering his brains over his gorgeous war bonnet of eagle feathers.

The momentum of the charge carried Jim through the thin line of warriors, but they were now between him and his friends the Crows. He wheeled, and before the Blackfeet could recover from their surprise, he hit the line again; this time from behind. Again an enemy of the Crow was knocked from his horse, his head shattered by another vicious blow.

Screaming a war cry equal to any ever uttered by his red brothers, Jim rejoined the Crows. He taunted them as being children and cowards and challenged them to stay with him on his next pass at the line of the hated Blackfeet. The Crows were shamed into staying with him on the next charge, but the Blackfeet broke and ran before the charging lines clashed. Twenty or more were picked off one by one as their horses gave out, or they were injured during the headlong dash over the broken ground of the mountain and hillsides.

The Crow party returned to their village with Jim as their acknowledged hero and new leader. He once more became the son-in-law of the village chief. This time he became genuinely fond of his Indian wife. He stayed with the Crows until, in due course, he became not only a full chief of his own village, but second in command of the whole Crow Nation. His eldest son, Black Panther, eventually became chief in his own right, but that, of course, was years after Jim had moved on to greener pastures.[2]

Beckwourth was too restless to stay long in one place. He had to see what lay on the other side of the mountain. It is not surprising, then, that we find him with Joe Walker (Walker Pass and Walker Lake) when he led one contingent of the Booneville company to the Pacific Coast in 1832. There Jim met Charles Dana, who was in Monterey at the time, and who mentions Jim Beckwourth as one of the Mountain Men he met in California.

Jim was not impressed with the young sailor, and certainly not with the fur trapping possibilities in the California mountains, but he was astonished by the thousands of beautiful Spanish horses running loose in the coastal hills; free for the taking.

To the Crow, as well as to the other Indians of the plains, horse stealing rated just slightly below the lifting of a scalp, and the Crow, who had taught Jim the art, were acknowledged the greatest horse thieves of them all. In California no one ever stole a horse. If one was needed, it was taken. The owner, if indeed it had an owner, would reason that it must have been required. He also knew that if he should ever need one, he could feel free to take it.

In early 1846, Jim was visiting his family in their village on the headwaters of the Powder River, when a neighboring chief came calling, primarily to brag about his great exploit in

---

[2]This episode was told by Jim to James Bonner who wrote Jim's biography. It also probably stretches the truth to the breaking point. But independent research has established that the basic facts are true.

stealing fifty horses from the Spanish in northern New Mexico. Jim laughed at this. He told his people that this had been just child's play and not worthy of a real Crow chief. He was, of course challenged to do better. In true Indian fashion, he strutted about boasting of what he could do. He would become the greatest horse thief the world had ever seen. In fact, if he could not lead a raid that would return with two thousand horses [more than was currently owned by the entire Crow Nation] he, the Chief called "Bloody Arm" would never return to the bright land of the rivers, of the mountains, and of the buffalo, again.

Leading a band of twenty braves, Beckwourth traveled down along the eastern shore of the Great Salt Lake. Averaging nearly fifty miles a day, they cut across south central Utah, southern Nevada, and on across the Mojave. When they arrived in the region of the Tehachapi Pass, they discarded their scrub stock of Indian ponies. They caught and broke mounts from the herds of fine California horses that were the direct descendants of the Arabian and the Andalusian war horses brought to the new world by Cortez and improved by the selective breeding practiced by the Mexican ranchers before they were allowed to escape to the hills.

Now being as well mounted as the Californians who were bound to pursue them in force, the Crows deployed over the brown hills east and west of Tejon, and on a prearranged signal, struck at day-break. They hazed the semi-wild horses from a half million acres into the funnel of the Tehachapi Pass and on east out into the Mojave Desert.

It took the Californians two full days to round up a posse of forty vaqueros and begin the pursuit. Beckwourth and his braves were hazing 5,000 head of semi-wild horses, so they made no effort to hide their trail, but pushed eastward as fast as possible. Even so, the California riders, each leading a spare horse, gained rapidly on the more slowly moving herd.

The Indians pushed the herd so hard that by the end of the second day the old and the infirm horses and every colt under one year of age had either died of overheating or from exhaustion, or had been left hopelessly behind to die in the vast wastelands of the Mojave. The "solid" stock still numbered more than 4,000 when they reached water and pasture thirty miles east of the present city of Barstow on the Mojave River.

They watered the horses and scattered them over the pasture land along the river, secure in the knowledge that none would stray out into the desert and be lost. Then they set a trap and ambushed the pursuit which they knew could not be far behind.

Within hours, the Californians did arrive. They were desperate for water after two days of hard riding. They were also sure that their revenge on the Indians would be swift and complete. Almost at once they found that they were not dealing with the mild mannered coastal Indians. Beckwourth scattered his men along the rim of a draw leading down from the high desert to the river bottom where the horses were grazing for all to see. Jim had his braves fire over the heads of the vaqueros, then rode alone up the draw to meet the trapped men. He assured the leader that his Indians were not after scalps; they were only interested in horses. Jim spoke enough Spanish to make his captives aware of how fortunate they were in having been taken captive by a "gentleman," even though he was indeed a horse thief. They agreed, and were most grateful when they were allowed to get to the water. They objected when their superb mounts were confiscated by the Indians and they were given scrub stock to ride across the hundred miles of waterless desert back to Tehachipi and water. Jim had only to have his men stand from the concealed hiding places and fire a few rounds into the air. The crack of the Kentucky rifles convinced the trapped men of their immediate peril. They suddenly became glad to ride the inferior mounts, sans saddle and bridle, back to California.

Jim's Indians rode the fine personal mounts of the Californians, outfitted with the heavy silver-trimmed saddles and bridles for a few hours only. Then they tired of this new uncomfortable way of sitting a horse. They stripped the ornamental silver off the leather, made necklaces of it, threw away the fine leather, and from there on rode their mounts bareback, or at best with just their blankets as a pad.

With all pursuit eliminated, Jim guided the horses over the "Old Spanish Trail" across the desert north of the Colorado River. They crossed that stream just above Glen Canyon at "The Crossing of the Fathers," continued on up the San Juan River past Ship Rock, and on to Santa Fe.

There, Jim heard that General Kearney was on his way west with a detachment intent on taking New Mexico and invading the northern Mexican Provinces, the war with Mexico having just been declared. He reasoned that the General would have lost many of his cavalry horses on the long march from Fort Levenworth to the Rocky Mountains, so he took 2,000 head of his California horses on east. He met the General south of Pueblo, Colorado. Exactly as Jim had reasoned, he was unable to move farther due to the inability of the farm raised horses from the midwest to live off the country as they crossed the semi-arid lands of the high plains. The remount officers gladly paid $50.00 per head for the desert wise horses from California.

Jim returned to New Mexico and invested his

money in a large ranch near Taos. Then he gave one thousand horses which had survived the murderous drive across the deserts and over the mountains, to his Indians as their part of the profits from the raid and joined them on their drive back to the Yellowstone in order that the Tribe would know that he was indeed the greatest horse thief who had ever lived! He then returned to his ranch on the headwaters of the Rio Grande River. He married a half Spanish girl, this time with the aid of the clergy, and settled down to a life as a respectable New Mexico rancher, raising fine horses, fine cattle, and he hoped, fine children.[3]

---

[3]Again most of the horse stealing episode was probably pure fiction. Jim stole the horses all right; but there were only 500, at least that was the number he delivered to General Kearney near Pueblo. Also the horses did not come from the Tejon area, they were rounded up on the great Cucumunga alluvial fan on the Los Angeles coastal plains. This followed General Castro's failure to pay Jim and his Mountain Men companions for their part in winning the battle of Cajunga Pass in the San Fernando Valley.

General Sutter of Sutter's Fort had joined the Governor from Monterey to help put down the rebellion in southern California. Several American settlers from the Sacramento Valley and from the San Francisco area made up the real fighting force of the Governor's army. When they rode into battle and found their horses being shot from under them, and heard the crack of the Mountain Men's rifles (they had been assured that they would face only smooth bore muskets); they lost all heart for the fight. Castro won the battle and the war.

The payment Jim and his men got was only an invitation to the great fiesta held on the Plaza

Destiny had other plans for the ex-slave boy from Missouri. The discovery of gold in California caused him to be practically drafted by the U.S. Army to guide Major Edward Beale (Camp Beale and Fort Tejon in California) in his successful attempt to build a wagon road from Santa Fe to Barstow via a route directly across northern Arizona.

His long experience in mountain travel guided Jim instinctively along the route through what is now Gallop, Flagstaff, Kingman, and across the Colorado at Needles. Then on to the Southern mines via Tehachapi Pass, and up the San Joaquin Valley. This is precisely the route followed by the famous Route 66, recently changed to Interstate 30.

Beckwourth mined for a very short time on the Stanislaus River, but the hard dirty work was not fitting for the proud men who had once trapped the beaver. Working all day in the icy waters of a beaver stream they would do, but grubbing for an ounce or two of gold was something the Mountain Men would not do. To them that was the work of squaws. So Jim left the river and started a store in the new town of Sonora. His experience trading with the Indians qualified him as a supplier of the miners in the rough community. He made money, but the quiet life in the store was not for him. He found a buyer in a miner who had struck it rich and was looking for a place to invest his gold. Then he returned to the life he knew best; that of exploring, and laying out new roads through the mountains.

[footnote 3, con't] in the village of Los Angeles. Nothing more. To collect the promised pay, Jim and his men rounded up the horses of Castro's friends, drove them east and sold them to Kearney.

He remembered that his friend Kit Carson had traveled up the eastern side of the high Sierra Mountains with Fremont in 1842. He had talked of the break from the granite range which rose sharply out of the Nevada desert for only a short distance past the blue lake called Tahoe. Then the Sierra ceased and the Cascade Range continued on to the north to the Columbia River.

Beckwourth reasoned, correctly, that where the two mountain ranges joined, there would be a break in the escarpment which guarded the entrance to California.

Carson had also told Jim that the Cascades were made up of a long chain of volcanic peaks. If that were true, then it followed that there must be low "saddle backs" between the peaks and, most likely, there would be a low pass where the two mountain systems joined.

Jim's reasoning was correct. He traveled north up the Sacramento Valley until he could see the outline of Mount Lassen on the horizon. This must be the first of the volcanoes Carson had told of; so he turned into the hills and followed the first large river that appeared to originate south of the snow-covered, rounded peak to its source.

Again Jim's instinct paid off. He had followed the Middle Fork of the Feather River into Sierra Valley. From there it was hardly a five hundred foot climb up a smooth plain covered with sage brush to the gentle slope down into Long's Valley and the flatlands of the Nevada desert.

On his first try, Jim found a pass through the mountains which was eventually to save hundreds of lives as the great rush to the gold fields continued. No longer would wagons have to be taken apart to be hauled up the face of cliffs to the summit of the Sierra, there to be reassembled and driven only a few miles, then be broken down once more, this time to be lowered by ropes down the face of another cliff on the western slope of the mountains.

Jim built a cabin combined with a storeroom for his trading post. This he located hard by the banks of the river, just where it turned from a gentle, sparkling mountain meadow stream, into a leaping, tumbling, raucous mountain torrent, carving a deep canyon through the mountains to the west.[4]

To complete the way to the gold fields, Jim built a road up to the long gentle slope of the ridge between the middle and the south forks of the Feather. This was to make the last thirty miles of the '49ers an easy down-hill pull, and to allow his own supplies to be hauled from the head of navigation at Marysville to his trading post rather than be packed in on the backs of mules. Then he sent two men east to the valley of the Humbolt River to intercept the wagon trains as they were about to begin the deadly crossing of the Humbolt Sink.

Jim had already established a reputation for finding new and easier routes, so when the captains of the wagon trains were told that Beckwourth's Pass was now open, they turned west, passed Pyramid Lake and followed Long's Valley to Hallelujah Junction, (well named—they had reached the land of gold!), and into Sierra Valley.

At Jim's trading post they found the facilities to repair their wagons from the long trip across the rocky roads and through the sands of the deserts. The livestock were almost dead from pulling their loads and from the lack of both

---

[4]The cabin/trading post is still there! The logs, hewn from ponderosa pine trees by Jim's own hands may still be seen. Until quite recently it was occupied by a single ranch hand who did his own cooking. On an electric range, yet! And there he spends the long winter evenings watching television!

water and grass. Beckwourth said it best. "They were so emaciated that the only way they could stand was to be held up by the tails, by the drivers."

Jim sold the emigrants flour and bacon to replenish their exhausted supplies, then he traded new and fresh horses and oxen for the stock which could go no farther. The "boot" was fifty dollars or more. Jim had learned the business of running a trading post in the Indian country where the mark-up was at least ten times the value of the goods at the nearest "rail head." For a time he followed this practice on the Feather River. The starving livestock were put out to pasture on the lush grasses along the river in Sierra Valley. Within six weeks they were "fat and sassy" once more; now, Jim traded them to the next wagon train for the usual "boot." He got rich fast!

Left alone, he would have become one of the great men of California history. But from the beginning, he had two strikes against him. He was a negro, and technically still a slave; and he by his own admission, was the greatest horse thief who had ever crossed the plains. [He was passing as a Crow Chief, and everyone knew that to the Indian, "horses were for stealing."] He still might have survived, had he not made a serious mistake.

Before he had completed the road down the ridge into Bidwell's Bar on the Feather River above Oroville, he approached the City Fathers of the port city of Marysville for assistance in building the necessary bridges, etc., on his road. He pointed out that his new road would bring tens of thousands of travelers through their town, and its docks would handle all the supplies going to the Northern Mines. They could compete successfully with the ports of Sacramento and Stockton.

The city agreed that they would pay him

$5,000 when he led the first wagon train coming over his pass into their city. Jim completed the road using his own money. Then he rode proudly at the head of a train of fifty wagons down the already well traveled road from Bidwell's Bar into Marysville. When a great cloud of dust welled up on the northern horizon, the Mayor of Marysville sent a rider out on a fast horse to see if indeed it was Beckwourth and the promised gold seekers.

The rider came dashing back waving his hat and shouting the glorious news that Marysville was now to become one of the great cities on the Pacific Coast. Immediately the mayor declared a holiday and led an impromptu parade down the main street. The saloon keepers passed out free drinks and the merchants handed out blasting powder for the explosions which must accompany every celebration. Unfortunately, some over-enthusiastic celebrant pushed caution aside. He set off a whole keg of gun powder! The resulting explosion scattered several burning buildings to the winds. The whole town burned down before the already intoxicated volunteer firemen could be assembled to man their hand-drawn, hand-powered pumps.

When Jim demanded his $5,000, the mayor was forced to explain that with the town destroyed, there was, nor would there ever be, any funds to pay the city's obligations.

Jim had been through this before. He shrugged and waited for a moonless night; then he rounded up about $10,000 worth of horses and cattle and drove them to his mountain retreat in Sierra Valley. Some he traded to the wagon trains as they came to his trading post. Most were held as breeding stock to develop a herd to feed on, and grow fat on, the lush grasses of his 50,000 acre ranch.

Again Jim might have survived had he not aroused the ire of the large mining companies, especially those who were diverting the rivers to glean the gold from the deep holes. He had hardly become established as a wealthy rancher, when he was approached by one of the workers who had been helping in the recovery of the large nuggets from the bottom of the diverted rivers. The worker had put a few of these nuggets in his pockets from time to time. Now he had several thousand dollars in gold which was unspendable in large amounts. He had already taken up a piece of land on the lower Yuba River. He needed cattle and horses to stock his range land. Would Jim trade him breeding stock for his raw gold, at $50.00 a head?

Jim would, and he did. How else could he make money in such large amounts so easily? He loaded some of the gold into his saddle bags and rode south to the Tejon area where he claimed he had stolen the horses in 1846. This time he paid for the cattle and horses he drove away.

He had hardly settled down to enjoy the easy profits he had made by dealing in the highgrade gold, when he was approached by yet another miner who also had empty range land and a poke full of stolen gold.

Once more Jim rode south with his saddle bags bulging with gold. This time he crossed over into the Los Angeles Basin where millions of cattle and horses still roamed the plains and the open hillsides, much as they had done before Marshall upset the scheme of things on the haciendas of California. This time he drove a thousand head north to restock his range.

The mining companies were not run by stupid men. They knew exactly what had been going on. After two years of prodding the local sheriffs to stop the larceny, they despaired of getting relief from the law enforcing agencies who, admittedly were afraid to venture into Beckwourth's domain to put a stop to his illegal activities. The companies took the law into their own hands.

They published the fact that Jim was no Indian Chief, but was instead, a "nigger slave" trying to pass in a white man's world. They approached a gang of murderous cutthroats who had been stealing cattle and horses from the ranches up and down the Sacramento Valley. They promised them immunity if they would ride up to the Sierra Valley and hang the "nigger"! They could also keep everything they could carry or drive away.

Jim had his spys out to warn him of just such a contingency. He also had two of the fastest mountain wise horses in California, corralled behind his trading post. Given a head start of five minutes, the man had yet to be born who could catch Jim Beckwourth on a cross country ride through the mountains and over the deserts. Jim had ten.

He rode east to the Black Fork of the Green River in south-western Wyoming. There he worked for a time tending store for Jim Bridger. When the U.S. Army took over that post as they waited for re-enforcements before their avowed purpose of invading the valley of the Great Salt Lake, to drive out the Mormon polygamists, he traveled still farther to the east and opened a small store on the outskirts of the new town of Denver.

Once more he married. At least he took a squaw into his bed, and used her other services to keep the store clean and to wait on the customers, most of whom were Indians.[5]

In 1876, after more than fifty years of wading in the icy waters of the beaver streams, of

---

[5]Before taking the squaw, Jim married a well known widow from Denver. He set her up in a large house built on his ranch on the South Platte River. The marriage did not last. She could not put up with Jim's Indian ways. She demanded and got the prosperous ranch when she testified that she had not known Jim was black.

battling Indians, and all too often white men, and of riding for days and even for months across the sun scorched deserts and over the snow covered mountains, Jim's rhumatism-wracked body became too much to bear. He sent for his oldest son, Black Panther, now the chief of his village on the Yellowstone River. Together they rode the plains back to his first love, the Indians. The ride which would have been completed in two weeks in the days of Jim's prime, with the nights spent in an Indian village after fifty miles in the saddle, feasting, dancing and bedding with the maiden of his choice; now took more than a full month, each night made as comfortable as possible by the son gathering dried prairie grasses to make into a thick mattress to soften the aches of the weary joints of the old warrior.

They arrived safely at the village, but Jim died the next day! Legend says that he was poisoned by the Indians who knew of his restlessness. They knew that if he were to leave them again he would never return. They believed that even in death his spirit would never return. They believed that even in death his spirit would be strong enough to bring back the days of glory to the Crow of the plains. The medicine of Chief Bloody Arm, the greatest of the Crow War Chiefs would drive all enemies, even the white pony soldiers, before the irresistible charge of their warriors.

Once more the tepees of the village could follow the buffalo herds, and squaws could prepare the hides for soft beds and to cover the poles of the lodges to keep out the bitter cold of the northern winters. Once again the bellies of the children would cease to show the bloat of starvation, and they would return once more to their games of hunting, horse stealing, and the taking of scalps.

They buried him with all the honors of a

great Crow War Chief. He was dressed in beaded deer skin made soft and pliable by being chewed by a dozen squaws. On his chest, he carried a gorgeously painted war shield, and on his head they placed a full headdress of eagle feathers, their greatest tribute to a tribal leader.

His body was carried to the crest of a hill from where his spirit could see once more the plains stretching a hundred miles to the eastern horizon; plains where he had once followed the buffalo on its annual migration from Texas to the Canadian border, often forming a solid mass of grazing beasts over the plains as far as the eye could see. They erected a burial platform and placed the body so that he could see not only the great plains, but could look down on the clear mountain stream meandering through the meadow below—the stream which was once his favorite beaver trapping creek.[6]

---

[6] The poisoning was not necessary. He arrived at the village tired and broken from the long ride. The Indians insisted on a great feast of Texas steer meat. Steers driven up the Chishome Trail and given to them by the army. Then they danced and sang as they had in the days when he had been in the prime of his manhood. At last he was allowed to retire to the lodge of his favorite Indian wife. Her demands, plus all the other demands on his weakened body, proved to be too much. He died in her arms!

CHAPTER XV

GOLD IN QUARTZ

Once the rains and the snows had sent the miners to shelter after the lush pickings of the summer of 1848, the idle men began to speculate as to the source of all that gold which they now knew to be scattered along the Sierra foothills from Fresno to the Feather Rivers.

Some envisioned a great volcanic explosion which threw the nuggets and the dust high into the air. The gold fell back to earth, scattering over an area of more than 1,200 square miles. Then as time wore on, it was supposed to have been concentrated in the stream gravels by the constant erosive and sorting actions of the streams of the western Sierra.

Others noticed that there was not much, if any, gold to be found above the 4,000 foot level in the mountains. They also noticed that the slates, limestones, serpentines, and the griesses were found below that elevation, while the granites, the diorites, and the granitic schists were above. They reasoned that there must have been some force deep within the earth which had squeezed the precious metal up through a crack which developed between the dissimilar rocks. It was then spread out by the running water which poured down over the sides of the mountains. To a slight extent their surmises were correct. The great quartz ledge, which often swelled to a thickness of forty feet or more, and a very small portion of which did carry gold in commercial quantities, had indeed been formed along a fracture line running from Mariposa to just south of Placerville. This ledge was more nearly on the

2,000 foot line, rather than much higher as the first miners had speculated, but it did contribute much of the gold found in the streams below. Upon the discovery of its existence, it promptly was named the "Mother Lode Ledge," and the 150 miles of foothill land it bisected was immediately named "The Mother Lode."

During the summers of 1848-49, the more observant miners must have noticed that the gold they found in the stream beds was always associated with white quartz pebbles and white sand. Many chunks of quartz must have been picked up which showed flecks of gold on the surface. Mostly they were smashed to release the metal. No one, however, associated these stray pieces with the real origin of the gold in the streams. Every miner was too intent on cleaning up his present claim, and joining the next stampede to the new diggings where, according to the latest rumors, the placer miners were walking away from claims which did not pay a dollar a pan.

Very often the claims were too conservative. The gold was there for the taking. According to General Beale, the man who was really responsible for the explorations of the Southwest credited to Fremont, and who built the wagon road from Santa Fe to Barstow, "many of the early arrivals in the gold fields panned as much as ten to fifteen thousand dollars in gold before the winter rains sent them to shelter." They were quick to learn the rudiments of placer mining and they were extremely lucky in their choices of locations to begin digging for their fortunes.

The first real knowledge of the source of the gold came early in 1850. Henry Meyers of Rough and Ready is supposed to have picked up a piece of quartz on the trail up Squirrel Creek to Boston Ravine. A small amount of gold was embedded in the quartz. Meyers was not sure that it was gold; so he took it to B.L. Lamarque, the saloon keeper at the Boston House in the Ravine. Lamarque was the authority on such matters in the Wolf Creek diggings. He examined it and pronounced it gold. He advised Meyers to crack it up and pan it. The amount of gold recovered was discouraging, so nothing more was done about the find. The honor of the discovery of gold in quartz goes to another, a man with the good sense to follow his observations.

In October, 1850, George McKnight[1] partially solved the mystery of the gold in the streams of California. George was working a small tributary which emptied into Wolf Creek at Boston Ravine. Because, at that time of the year, there was no water in Rhode Island Ravine where he was working, he was laboriously carrying the scrapings from the bedrock of his claim across the shoulder of a small hill and on down to Wolf Creek where water for panning was available. His claim was rather rich, but with all the work involved, it was hardly worth the effort for the five or six dollars a day he could recover. It needs no great imagination to see him kick in disgust a small rock which was lying on the trail he and others had worn along the hillside.

Mrs O'Leary's cow kicked over the lantern in a barn in Chicago which started a fire that made it the most expensive kick of all time. McKnight's kick was of equal historical importance. It was the richest blow ever struck by the toe of a man's boot!

If we give McKnight credit for only the gold dug out of the fifty acres on the slopes of Gold Hill, we find that $3,000,000 was dug out of the first few feet of the surface before Christmas of that year. No one knows how much was eventually

---

[1]The bronze plaque on the monument on Gold Hill says that "George Knight" made the discovery. This is an error.

taken from the 10,000 feet below by the deep rock miners who followed the first prospectors. Probably a minimum of fifty times as much.

Within a radius of two miles of the discovery point, a hundred mines took out from one to ten million dollars before they were worked out or were consolidated with one of their neighbors. In addition, the Empire, the North Star, the Pennsylvania, the Allison Ranch, the Eureka, the Idaho-Maryland, and the Brunswick Mines each brought up quartz worth from fifty to two hundred million! And these sums are based on the price the U.S. Mint was paying for the gold when it was produced—$20.00 to $35.00 per ounce.

McKnight and his neighbors dug out so much gold with so little effort that they rushed to the local stores to turn it into tangible goods, or into silver before it lost its monetary value. They reasoned that the ledges they had found would continue on down to the molten rock below the earth's cool crust, and if the percentage of gold the quartz contained was constant, the amount would be so great that its value would be lower than that of silver.

McKnight's kick sent the rock rolling along the path ahead. As it rolled over and over, the rays of the setting sun reflected from a metal on its surface. That strange phenomenon distracted his attention from the pain in his toe. He cursed a little, then he picked up the rock to see what had made it reflect the light in such a strange way. He was startled to find that this rock seemed to weigh considerably more than a piece of quartz that size should weigh. He looked again. Where it was freshly broken, it showed a metal with a bright copperish color. George needed no acid nor scales, as had Sutter and Marshall two years before, to tell him that this was GOLD!

He dropped the sack of gravel by the side of the path, and forgetting the thinness of his boots and the pain in his toe, he ran down to the spot on the creek bank where he had left his gold pan. He placed a flat rock in the pan to act as an anvil and broke the prize piece of quartz with a hammer, taking care not to let the chips fly away. The quartz had been near the surface for a very long time. It was pitted by weathering so that only a little pounding reduced it to pieces small enough to release the embedded gold. Now it was a simple matter to pan it, exactly as he would have panned the black sand which had collected back of the riffles of his rocker.

The crushed quartz panned off easily, and McKnight gasped as he saw the ribbon of gold that collected in the crease on the lower side of his pan. He, nor anyone else had ever seen anything to compare with this, but he had had enough experience panning gold to know that in just a few minutes he had recovered more gold from that two pound rock than he had been able to glean from the stream bed in a whole week of the hardest kind of labor.

As soon as he had cooked and eaten his supper, George hiked down to the Boston House for his usual drink with the boys, then he did a most unusual thing. He bought a bottle of the bar's best whiskey to take home. "For his cough," he said. The bar man looked long and hard at the irregular chunks of gold he was given in payment. It looked like nothing he had ever seen before. The pieces showed none of the rounded, water-worn formation of the gold that had been carried along with the gravel on the stream beds. However it looked like gold and weighed up like gold; and anyway it might not be too healthy to question these tough miners about the quality of their money. He accepted it without a word and put it aside for a more careful examination later.

McKnight went home and, after a couple of stiff jolts from the bottle, went to bed, but not to sleep. The whiskey acted as a depressant to dull his senses, but he was so keyed up by his good fortune that sleep would not come. At last, in his half stupor, from the alcohol and from fitful sleep, he visualized again and again the glint of the metal as the rock bounced along the trail. Gradually it occurred to him that this might not have been just a random piece such as those which were sometimes picked out of the stream gravel, but that there might be more rock just like it at the same location. He made up his mind that as soon as it was light enough to work, he would return to the spot and do some digging. With that problem settled, he dozed off and slept fitfully until dawn sent some light filtering in around the blanket which acted as the door to his pine pole and brush cabin.

George stirred the coals from the supper fire and quickly heated some coffee, he downed a cupful of the hot liquid and ate a couple of flint-like sea biscuits, then he hiked back up the trail he had run down the evening before. The spot was easy to find. His pack sack filled with gravel was lying exactly where he had let it fall.

A quick inspection showed no more of the white rock with the flecks of gold. He hiked on up to his diggings in Rhode Island Ravine, picked up his mining tools and retreated to the point on the path he wanted to prospect. At the point on the trail where he had kicked at the small rock, he began to dig a trench about a foot wide. His pick struck rock almost immediately. Each piece that he separated from the red clay of the hillside he examined carefully to see if by chance it matched the "treasure" he had found the night before. Several of the other miners came along on their way to their claims above. They paused, scratched their heads questioningly, and asked, "What the 'ell was goin' on?"

McKnight was not anxious for them to know

what he really was doing. First, if he did indeed have something good, he was not about to share the good news until he had "got his." Then too, he was afraid that the piece he had picked up was just another bit of "float rock" containing gold. If that was the case, he knew that he would never hear the end of his "pipe dream" if the boys at the saloon found his secret.

He told them that the evening before he had thought that he had seen indications of gravel left by the stream when it had been busy cutting away the hillside a million years before. Now he was trying to locate it with a cross cut along the path. That explanation satisfied them; they went on their way, and he went back to digging.

A few feet of trenching led him back to the exact spot where his toe kicked the rock loose. A few shovelfuls of dirt thrown to one side, a little picking away of the dirt, and a quartz ledge about four inches wide was clearly defined.

McKnight had seen quartz ledges as they cut across the bare granite in the high Sierra and where placer miners had exposed them as they washed the gravel down to bedrock in their efforts to get at the fine gold in the cracks and crevices. But this was a very special kind of ledge. Instead of being solid white quartz which "rang" with a special sound when struck by a pick, this "stringer" or narrow ledge, was stained brownish-red by the iron oxide it contained. Also, instead of being solid, it was partially filled with cavities, some of which were filled with GOLD! The pieces he broke out matched exactly the quartz he had picked up the night before.

McKnight dropped his pick. He grabbed a handful of the broken "HIGHGRADE." He trembled so violently that he was forced to sit on the bank of the trench he had cut. The implication of what he had discovered was not clear to him, but he knew that if the ledge he had found continued any distance, he had enough gold in sight to never again have to break his back cleaning gravel from around the pine stumps and the boulders on the stream banks, or on the stream beds. Never again would he have to court the chills and the fevers and eventually the rheumatism which must result from days on end of working in the cold water as he grubbed for the few dollars a day he could hope to recover.

After a few minutes he recovered enough to go back to work uncovering more of the gold-bearing rock. He followed the ledge several feet in each direction from the discovery point, and dug down to a depth of four feet at one spot to be sure that the quartz did continue down into the semi-solid rock below. It did! A quick mental bit of arithmetic made him realize that in less than three hours work, he had uncovered enough gold-bearing quartz to give up a minimum of $10,000!

When the full realization of what he had found struck him, McKnight went into a state of shock. First he wanted to laugh, then he found himself crying. In spite of his better judgement, he felt that he must share his good fortune with his friends. He knew that he should cover up his find and work it secretly. He was well aware that a major strike could lead to a stampede to a mining camp. Too often the discoverer of a new diggings was pushed aside by the greed of the new arrivals. Claim jumping and often murder were the accepted methods used by the desperate men who had arrived too late to get in on the easy diggings. In desperation they would commit acts which, under ordinary circumstances, would be farthest from their thoughts.

But McKnight threw caution to the wind. He grabbed a double handful of the rock and ran up the creek to where his neighbors were working. They must have thought him mad as they saw him running across the hillside shouting at the top of his voice: "Boys, I've got it! I've found the source of all the gold in these Californi' hills!"

McKnight had indeed found the key to millions, but never in his wildest imagination could he foresee the problems his discovery would lead to. All he knew was that he had a small fortune in his hands, and there seemed to be an unlimited supply where that came from. No wonder his face was flushed, his eyes wild, and his voice hoarse from yells as he charged into the '49ers along the creek bottom.

Some of the red shirted workers were skeptical. They only laughed at his claims, but such were the times that mostly they were ready to believe the wildest of stories, especially if there were the smallest bits of tangible evidence, and of more importance, there was a chance that they might be on the spot and be one of the first to stake a claim. They dropped their tools and gathered around McKnight demanding to see the evidence. They wanted to "heft" the rock again and again, for only when they had the stuff in their hands could they believe that it was as heavy as it really was.

After McKnight got his breath and became coherent, he explained how he had found the rock and how he had gotten the gold out of it. He reasoned that now that he was really rich he could afford to be generous. He invited them to come along and he would show them how to get out of the muck of the stream beds and how to make their fortunes at the same time.

As the miners walked back along the trail to McKnights diggings, they were already working out some method of dividing the wealth. Until now, the crude mining laws allowed each man to claim a certain number of feet along the channel of a stream, and up to the crest of the first rise on either side, the width being dependent upon the richnesss of the gravel closest to the running water. There were already some exceptions. On the limestone flats at Columbia, on the slopes above the rich diggings on Deer Creek at Nevada City, and on the gravel flats at Yreka, the claims were limited to ten feet square, one hundred square feet. At Grass Valley at the time, the rule was that a man could claim fifty feet.

In all fairness, we must give credit to the restraint of McKnight's friends. Lacking any method of staking "lode" claims, they allowed him fifty feet along the creek below and let it run to the crest of the hill above in such a manner as to include the discovery point.

Almost immediately this method of staking quartz claims was amended. From that time on, the discoverer was allowed to take fifty feet on either side of the discovery "hole," and five hundred feet along the "strike" of the ledge. A few years later, the law of Caliifornia made it possible for the claimant to take 500 feet in width and 1,500 feet along the strike (length). Also a man might take as many claims as he wishes, so long as the ground is open for entry, and he thinks it is prudent to dig a ten foot shaft, and do $100.00 worth of work on each claim during each calendar year (assessment work).

The discovery that gold was found in the interior of solid pieces of quartz rocks did much to solve the mystery of the gold in the streams of California. The running water had "mined" the quartz. It had then worked its way down the hillsides to the running water below. There, due to the crushing and grinding of the rocks and gravels being carried along by rushing water, the gold was released from its imprisonment. It then was carried downstream, and as it moved along it became rounded and smooth; the typical form of placer gold.

Next the question arose as to how the gold

got in the quartz in the first place. The first and the most widely held explanation was that the gold and the quartz existed as liquids deep under the crust of the earth. Great cracks or fissures developed in the overlaying rocks. Pressures squeezed the quartz up through these openings; sometimes all the way to the surface, but not necessarily so. At the same time an infinitesimal amount of gold was also forced into the cracks and up towards the surface. During the squeezing and pushing, the gold mixed with the quartz, but rather than being mixed uniformly, it tended to stay in bunches. Hence it occurred in pockets or in chimneys.

The above explanation served until it was found that in very rare cases gold did not remain "free." Although it is supposed to be inert, it actually will combine with tellurium to form a telluride of gold. But even before that discovery was made, it was found that gold is often combined in the crystals of iron sulfide, a common compound in rocks the world over. Almost never, however, is it found in the pyrites [fools gold] other than when that compound is found in almost pure quartz, and almost never then unless there is some free gold in the quartz.

The explanation which seems to fit best is as follows: Deep under the surface of the earth there were pools of super-heated water. The temperature might go as high as several hundred thousand degrees centigrade. Under the pressure of millions of pounds per square inch, even the water at those almost unbelievable temperatures could not change to steam. It must remain in the liquid state.

Again, under those pressures and temperatures, many compounds and many elements will go into solution exactly as sugar will dissolve in hot tea. This fact is particularly true in the case of quartz, iron sulfide, and the royal metals.

With this in mind it is easy to see that it was not liquid quartz which was forced up through the fissures; rather, it was the hot water. As it made its way towards the surface the pressure must be lessened, it also follows that the temperature must fall in direct proportion to the drop in pressure. At some depth the temperature reached the point where the dissolved elements and compounds could no longer be held in solution. The quartz was "frozen" out of solution. In isolated places the frozen quartz trapped some of the solution, keeping it from being pushed higher through the crack in the earth's crust. Eventually, it too lost heat until the other compounds and elements which had been in solution were forced to solidify. If gold was the main element involved, free gold was deposited. If the compound of iron and sulfur was present, the gold could be precipitated along with the "pyrite," usually in microscopic particles in the interior of the iron sulfide crystals.

It was many years after McKnight found his quartz mine before these facts were known. As a result millions of dollars of gold closely associated with the sulfides were allowed to run down the creeks below the stamp mills to be lost forever. In a few cases, the assay of the quartz showed that it contained much more gold than was recovered. A second assay showed that the lost gold was still in the tailings. Some mills were able to store the tailings to be worked later. Many of the ponds containing the gold bearing sands lost in the early milling processes stood untouched until the Depression years, then some unemployed mining engineers installed semi-modern recovery machinery and took the sulfides out of the sand to be sent to a smeltery.

McKnight and his neighbors were satisfied with the free gold they found in the quartz they mined from the top few feet of "Gold Hill." The partners, Huff, Hugunin, and Crackling had been gathering materials to build a cabin for protection from the storms of the coming winter. They dug out a flat spot on the side of the hill just above Boston Ravine. In doing so, they found about two tons of a flat white rock about six inches in thickness. This they piled, intending it to be used in the construction of a fireplace.

When McKnight made his discovery they had second thoughts. They examined their white rock. It contained gold in visible amounts. They worked until the week before Christmas down by Wolf Creek grinding and panning their ore. Then they hiked down to Marysville and took the river steamer to San Francisco. At the U.S. Mint, they were given a draft on the U.S. Treasury in the amount of $30,000! Not bad for 10 weeks' work out of the rains of winter, under a shed they had built down by the creek.

REDUCTION GEAR, CORNISH PUMP
NORTH STAR POWERHOUSW

CHAPTER XVI

MINING CLAIMS

McKnight's find of gold in quartz on Gold Hill disrupted the new mining industry in California. The news of the strike spread from the Merced to the Feather Rivers with a speed only to be equaled by the gossip of a scandal in an Iowa town. Every miner was sure that, now that he knew that gold came out of the quartz ledges, his fortune was made. Each placer miner remembered that he had seen the white ledges outcropping along the banks of the streams he had mined. The ledges were there, but most of the men who abandoned their rockers and their sluice boxes to rush to the place where the ledges were seen were doomed to disappointment. All too often they were to find that after drilling and blasting, sinking shafts and tunnelling into the hillsides to get at the quartz, no gold was to be found in the quartz they had worked so hard to get.

There are literally millions of quartz ledges intruding into the igneous rocks of the high Sierra, and even more in the rocks of the foothill country along the Mother Lode and the Northern Mines. In all of these, probably less than one-tenth of one percent carry metallic minerals of any kind, and of those that do, less than one percent has minerals in sufficient quantities to justify the cost of mining.

In many ways the failure of the quartz miners was more tragic than that of the men who failed to find gold in the stream gravels. The placer miner found out very quickly if there was gold in the place he had chosen to work. If he found none, he was able to move on to a more likely spot with little bother and at almost no cost. The quartz miner might work for months or even for years before being convinced that the ledge he had been following was worthless.

The person who has never mined is usually at a loss to understand why the prospector will work so long on what eventually proves to be a hopeless cause. Of course he was foolish, but every miner is sure that "just one more round" will open up an ore shoot and the blast will expose the ore which will make all his dreams come true.

Many stories of wasted lives and broken dreams still circulate through the mining country. Most of them have some basis of truth, but usually they have been spiced up by the vivid imagination of the tellers.

The story of Michael Brannen, however, cannot be improved upon, and the physical proof of the tragedy is in the Greenwood Cemetery in Grass Valley for all to see.

Brannen was sent out from New York by a group of investors to develop a quartz ledge which had outcropped on the hillside immediately across Wolf Creek from Gold Hill where McKnight had made the discovery of gold in quartz. The ledge had, like most of the ledges in the area, produced some gold near the surface, and had in spots given up some pieces of very rich rock. On the strength of this history, the New York group had bought the mining claim, and had raised money to begin further development. Men were getting rich taking quartz from identical ledges lying within a few hundred yards. There seemed to be no reason why this particular ledge should not do as well.

Within months after Brannen began working the ledge, some gold was found. He sent east for his family secure in the belief that his fortune was made, and that his children would be able to grow to maturity in this beautiful little valley in the Sierra foothills.

The small pay shoot was soon worked out. The main shaft was deepened, the drifts were extended, and cross cuts were driven to tap adjoining ledges. Some gold was found, but not enough to show a profit, and often Brannen was forced to send east for more money to continue mining. Funds became harder and harder to raise even though Brannen wrote glowing accounts of what the neighboring mines were doing, and how he was sure that given just a little more time and given the money to hire miners and to pay for drill steel and blasting powder, everyone involved would soon become rich.

After three years of begging money from the eastern investors, and of one disappointment after another, the letter arrived from New York telling Brannen to close the mine. There would be no more money!

Brannen refused to quit. He registered the mining claim in his own name. He borrowed money, putting up the mine and his own house as security, and continued to follow the quartz ledge. Still no more gold.

At last he became depressed by his failure when all around him men were getting rich with far less work on ledges which did not seem as likely to produce gold as did his; and, after he had reached the limits of borrowing, and his family was actually suffering for lack of food, his depression became so great that he lost all reason. With his last money he sent to Sacramento for a bottle of prussic acid. When that arrived he laced the last of the breakfast food in the house with the poison and fed it to his wife and their three children. He ate nothing but proceeded to the mine as if going to work.

At the mine he loaded the last of his blasting powder into the holes he had already drilled at the end of the drift along the main ledge. Then he lighted the fuses in the proper order for the most effective breaking action, and calmly seated himself on the empty powder keg awaiting the blast.

The neighbors found the dead family and came to the mine with the tragic news. They found his mangled body under tons of broken quartz. The face of the drift now showed a two foot ledge of quartz literally sprinkled with bright and shining free gold!

The people of the town buried the family in a row of graves. They erected five headstones, each bearing the date of 1858, and as was the custom at the time, they had the five graves enclosed with a "fancy" wrought iron fence. [It remains there today for the visitors' inspection.]

A second version of the story, and one much more apt to be believed by the cynical members of our society, goes as follows.

Brannen did operate the mine until all hopes of meeting his debts to the local banker had been lost. He borrowed the last of the grocery money from his wife's cookie jar to buy enough powder to set off one last blast in the mine. "Glory be to God!" the blast brought down powder boxes full of "picture rock." A fortune lay at his feet!

He grabbed a double hand full of the highgrade and rushed to town and into the office of the president of the bank which held the mortgages on his home and on the mine. He showed the banker the rich rock and asked for an extention on his mortgages until he could have the ore crushed and the gold it contained extracted.

His "friend," the banker, congratulated him and told him to come back the following morning. He would have the proper papers drawn up for the necessary signatures. As soon as the banker saw that Brannen was busy setting up the drinks for the boys in the "Owl Tavern," he saddled his horse and galloped all the way to Nevada City. There he had the sheriff execute the foreclosure clauses of the mortgages. He now owned the mine!

The shock of finding out the depths of perfidy of which man is capable was too much for Brannen. With the highgrade he had carried away from the mine he sent for the poison. Then he bought a full keg of powder and carried it into the mine. He was determined that if he could not have the wealth he had found, he would make it as costly as possible for the banker to recover it. He placed the powder in a place where the blast would bring down thousands of tons of rock to block the drifts. Then he seated himself on the keg and lighted the fuse!

Regardless of which story is true, it is a known fact that after the death of the Brannen family, the mine was worked with some success for a number of years. At a depth of about 1,000 feet, the pay shoots were lost and the mine closed. In 1928, the Newmont Mining Corporation purchased all the mines which had been operating on the Empire Ledge which extended from near the top of Osborn Hill across Wolf Creek and well beyond Lafayette Hill to the west. All the satellite mines were closed and the mining carried on through the shafts of the Empire and the North Star mines. The underground workings were connected and drifts were run in all directions from the deepest parts of the mines. One drift from the bottom of the North Star Mine cut through the "virgin" ground some 8,000 feet below the last workings of the old Brannen Mine. The drift cut a large quartz ledge which showed some gold. Miners were put to work breaking out the ore and sending it to the mill. The ledge proved to be very rich in places, and nearly worthless in others, but the ledge was followed upward by "stoping" until it broke out into workings more than 70 years old. Following farther, the miners were able to clear out the old tunnel of the Brannen Mine! How many millions were eventually recovered from the

Brannen ledge is anyone's guess.

There appears to be something suicidal about the deep quartz miners. Buckley Wells, the superintendent of the Idaho-Maryland Mine during the time that the Whitney interests were trying to find the long-lost faulted ledge, was unable to withstand the shock when Mike McGuire, working for Errol MacBoyle tapped the vast treasure and proved that he (Wells) had overlooked a fortune. He traveled to San Francisco and registered at the Clift Hotel. Then after brooding for a few more days, he put a bullet through his own head!

In December of 1934, in the Jones Memorial Hospital in Grass Valley were several patients undergoing physical and psychiatric treatment for the damage done by the breathing of too much oxide of nitrogen generated by a great underground blast of dynamite.

It seemed that a local miner had just lost his girl to a better looking rival. Somehow he got the keys to the underground powder magazine on the 5,000 foot level of the Empire Mine. Surrounded by a hundred boxes of 60% blasting gelatine (dynamite), he set off the charge. The resulting explosion did no great damage to the mine or to the other miners. The magazine was located well away from the working areas, and the rock in all the area under Ophir Hill is so solid that the blast did no more than cause a minor earthquake.

The love-crazed miner was, of course, disintegrated. All underground workers were ordered out of the mine, but the powder smoke spread so rapidly that several were caught in the deadly fumes. Two died, and several needed hospitalization for a long period to rid their systems of the toxic fumes, and to undo the psychological damage done by being trapped two miles below the surface of the earth with apparently no means of escape!

The uncertainties of the mining laws during the first few years of quartz mining led to any number of disputes, lawsuits, and undoubtedly to many murders. All kinds of disputes could, and usually did arise. For example, Ted Connolly made a nice strike on a ten inch ledge near Spring Flat. He staked the claim and recorded it according to the local mining laws. Then he found that the pay shoot sloped sharply towards the end line of his claim. By the time he had reached the depth of fifty vertical feet, he reached the end of his claim.

Meanwhile, Sam Jones had staked the land adjacent to the end line of Ted's claim. He worked night and day sinking a vertical shaft just within the end line of his property. He knew that the shaft would intercept the ledge being worked by Connolly at a depth of not more than fifty feet. It is not hard to imagine what would happen when Connelly broke into Jones' workings and found that his neighbor had walked off with the highgrade ore which he, Connelly, considered his by right of discovery.

Of course Ted would have known what had been going on by the activities of his neighbor; so the fighting would usually break out before contact was made underground. The mining laws were not too well defined, and in any case, then as today, the party with the most money or with the most friends around the courthouse, almost always won when lawsuits were filed; the grieved party was very apt to take the law into his own hands. In this case, Jones might well be looking down the barrel of a Colt forty-four when he came up the ladder out of his shaft. Then if he did not take the warning seriously, he probably would be met by a blast from a shotgun the next time he came up. Who was to say that Connolly had not been just trying to pick off a jack rabbit for dinner when Jones unfortunately popped his head up just at the wrong time? Or the drastic action could be the other way around. Jones could see trouble coming so he would drop a half keg of powder fitted with a short fuse down Ted's shaft. Poor Connolly! His ladder must have "broke" as he was climbing to safety after having set his charges!

Another source of trouble was very similar to that just described. But instead of the chimney of rich ore slanting towards the end line of the claim, it stayed perpendicular to the out-cropping of the ledge at the surface. In this case the whole ledge might slant towards the side line of the claim. This is called "the dip of the vein." In this case, Jones could sink a shaft just beyond the sideline of Ted's claim. Again he would intercept the pay shoot before it could be worked out. Such apparent injustices often led to serious feuds.

That the mining shafts were that close together can easily be verified by walking through the woods on McGuire Flat. This is just north of Osborn Hill, and only a half mile east of the Empire Mine. Even today it is not safe to walk through this area at night without staying strictly to the roads or to the well-traveled trails. Some of the earliest mining shafts are still open to a depth of fifty feet or more. By County ordinance, they must be either filled or fenced. But all too often neither has been done. Even the land owners may not know that they are there, or after twenty years or more, the fences may have rotted and have fallen down and have not been replaced. Also, many of the old shafts were once used as garbage pits by the local residents. When they were completely filled they were covered with brush. As the garbage decayed it settled so that a fifty foot shaft might have settled as much as twenty-five feet. This now became a perfect pitfall.

As late as 1968, the rod man of a survey

party was backing through some low brush to get to the position as directed by the transit man. A sixth sense caused him to look behind before taking one more step to the rear. That one additional step would have plunged him into an old shaft of unknown depth. Its collar had been encased in masonry so that the shaft could be used as a well by a family living near by. The abandoned well had long been forgotten, although it was within fifteen feet of a well-traveled road.

Some of the early editions of the Grass Valley Union, and other Nevada County newspapers, give rather vivid accounts of the so-called "Osborn Hill War." The shooting took place in the middle of the 1850's. A group of miners tried to take one of their neighbor's claims by force. A local dentist by the name of McMurty, a brother of the claim owner, seemed to have been the only one to fire a shot in repelling the claim jumpers. He must have been a cool character for although badly wounded by a blast from a shotgun, he fired six shots from an old "Navy revolver." His fire was most effective. He killed five of the overly ambitious miners and wounded a sixth! How many more were killed "accidentally" as the feuds wore on, no one will ever know. Enough, however, to give substance to the belief that gold is indeed the root of many of man's woes.

The miners were practical men. They soon remedied some of the real injustices of the early rules covering the locations of claims. First the rule limiting a miner to a single claim was changed. First one additional claim was allowed, then that rule was expanded to five in order that close-by neighbors could be barred from taking advantage of the work already done on the original claim in any newly discovered mining district. Following this change, any prudent miner would always stake additional claims along the "strike" of the discovery ledge [the direction it ran, north and south, or east and west] to protect his end lines. He would also stake claims on either side of the original claim to protect himself if the pay shoot pitched sharply towards either side. In taking these precautions many lawsuits and many actual battles were averted.

The second innovation in the structure of the mining laws, and the one which still delights the heart of every mining attorney, states that a miner may follow a vein on all its dips, spurs, and angles as far as he likes, even if this makes it necessary for him to go under his neighbor's claim, providing that the vein "apexes," that is, it comes to the surface on the first miner's claim.

On paper the Apex law makes sense, but underground so many things can happen that even the best geologist can not predict what may happen a single foot beyond the end of a drift, or a foot above the workings in a stope. Many rich ledges turned into blank quartz after passing a stringer coming in from either the "hanging" or the "foot wall." The stringer, if followed, might intersect another ledge which apexed on an adjoining claim.

The owner of the second claim might be justified in his belief that his was the true apex, and that therefore he was the real owner of a claim which his neighbor had mined illegally. The neighbor would be just as sure that his was the real apex, and that therefore he was entitled to follow the second ledge on all its dips, angles, and spurs.

To make matters even worse, a third property owner might believe with some justification, that both the controversial ledges were in fact but extensions of his large ledge which he was mining at a much lower level! Of such things, good lawsuits were made; suits which were often carried on until the lawyers owned all the mines involved.

One of President Lincoln's last acts was to sign the Homestead Law. This was intended to be a reward for the soldiers who had served during the Civil War. All others were also included in the opportunity of getting title to 160 acres of land by living on it for three years. The title so acquired included the mineral rights, and superseded any extralateral rights the adjoining mining companies might have claimed. Many men who owned and lived on mining claims immediately filed homestead papers, and in due time owned their mining lands and their homes. The minerals under their homesteaded land, if any, now could not be touched by a neighboring miner.

The little city of Grass Valley was built in the center of a highly mineralized area. Immediately the Homestead Act was passed, the home owners filed on the land on which their houses sat. To forestall a rash of lawsuits, the Empire Mine bought the mineral rights under the whole town. The owners of some well located lots were paid more than the cost of the house they occupied for the right of the company to dig under their house. The only restriction being that no mining could take place within 100 feet of the surface.

The purchase turned out to be a good investment for the mining company. During the depression years it was proven that many miners would leave their homes, walk the mile or so to the collar of the Empire Mine, board a "down skip" to the 2,000 foot level, walk a half mile to another shaft, and descend another 2,000 feet, and walk to the stopes where they were assigned, and find that they were working directly below their own home; the one they had left only an hour before. They might be removing quartz which carried thousands of dollars per ton, but their father or their grandfather had sold their "birthright" for a hundred dollars or so, many years before.

When the mining companies began fighting among themselves as to the ownership of certain ledges, the lawyers had a field day. In 1927, Errol MacBoyle succeeded in "making" a mine out of the long abandoned Idaho-Maryland property. During the many years that the Idaho-Maryland ledge had been "lost," the Empire Mine pretty much controlled the town and the mining district of Grass Valley. It had been able to buy up all the mining property which seemed to have prospects of producing gold.

The strike in the Idaho-Maryland, in 1927, put a young dynamic mining company in the field. This made for some uncomfortable situations. The local management of the Empire was put on the "pan" for failing to acquire the Idaho property when it could have been had for little more than the current taxes. It was, therefore, rather antagonistic when the Idaho began buying other mining property, especially property which lay on the opposite side of the Empire and could very well contain veins which had apexed on their property. They saw trouble coming. The Boundry Mine was one of the properties in question. The Idaho began mining on the Boundry's main ledge. The Empire sued, claiming that the ledge was in fact but a spur of the Empire's main vein.

The Idaho offered to settle for $5,000,000, i.e., sell the mine to the Empire. The Empire claimed "blackmail" and the lawyers got rich. The court battle dragged on until the Government closed most of the mines during World War II. After the War, both companies decided that the possibilities of profits from the Boundry Mine did not justify more expenditures. The Idaho-Maryland made no effort to reopen the mine.

Lying one-half mile south and east of the Empire Mine's main shaft and directly between that mine's property and the Idaho-Maryland is the 160 acres worked as a placer mine from 1849 to 1895 and patented as a homestead in 1870 by Michael McGuire, Sr. In addition to the agricultural ground there were two mineral claims which had never been patented. In order to keep the title to these, assessment work had to be done on each claim each year. These two claims lay immediately east of the Empire property and separated it from the McGuire patented land.

During the summer of 1939, the author was doing the assessment work for the McGuires on the claims in question. In doing assessment work it is understood that no one expects any useful work in the sense of development of the mine to be done. Any values which might be on the claims would be hundreds of feet below the surface. Consequently, the man doing the assessment work starts digging a trench, throwing out enough dirt in the cool of the morning to make a showing, and spends the afternoon taking a nice long siesta in the shade of a sugar pine tree. The owners of the claims are not particularly interested in the amount of work done. They have only to show on their books that they paid out $100.00 on the claim and file an affidavit with the County Clerk to that effect.

One hot July afternoon, I was sitting on the edge of a trench I had dug that morning, watching the activities of a hill of ants, when just at three thirty, the blasting time at the Empire Mine, the ground jumped up and slapped me right in the seat of the pants! I thought at first that there had been an earthquake, but when eleven more shocks came at five second intervals, I realized that the Empire was blasting in a drift or in a stope right under where I was sitting, and not too deep at that.

That evening, I told Mike McGuire, Jr., who was managing the property for the McGuire family, what had taken place. The next afternoon, just before blasting time in the Empire, Mike joined me. Together we sat on the bank of the trench. This time we both were shaken up by the force of the blasts below. On the third day, Errol MacBoyle from the Idaho-Maryland Mine, sat with us. After the third hole was shot, and we all bounced off the edge of the trench to better absorb the shock, just as a rider will rise in the stirrups to reduce the shock of the saddle on a "hard riding" horse; MacBoyle turned to Mike and said, "OK, Mike, you win. Let's go up to the Title Company and draw up the papers."

The McGuire estate had been trying to sell the "ranch" to either the Empire or to the Idaho-Maryland for many years. They had asked $100,000 for both the surface and the mineral rights. MacBoyle thought that the surface had little if any value. He was willing to buy only the mineral rights. For those he was willing to pay $75,000.

Once more the irony of fate stepped in. The Idaho-Maryland Mine, which had been paying from fifty to one hundred percent dividends PER MONTH [provided the stock had been purchased in 1927 or '28, for from five to ten cents per share] began to lose money when World War II restricted gold mining. MacBoyle died in the late 1940's. The mine went into bankruptcy and was sold under the sheriff's hammer in 1962 for $50,000. This included the 1,200 plus acres of surface rights, another thousand acres of mineral rights, and the few buildings which remained.

Meanwhile the surface of the land, still owned by the McGuire family, had increased in value due to the migration to the hills by the people who wished to get away from the smog and the crime of the large cities. In 1967, it actually sold for more than $200,000! [I know exactly what the purchase price was—I bought it, to be subdivided.]

The history of each gold mine in California is a story unto itself. Each one had its start due to the original discovery. Some were found as was McKnight's ledge on Gold Hill. Others were found by examining the ledges exposed during placer mining in the stream channels, or by prospecting the quartz ledges exposed by natural erosion of the hillsides. Usually the finding of a paying ledge was not that simple.

In the Grass Valley area most of the mines were found by one of two methods. The first consisted of cutting a trench down to bedrock. The prospector knew that the base rock in that area was "shot" full of quartz ledges and stringers. It was only a matter of time and of luck until a trench, if dug long enough, would intersect one of these quartz outcroppings. Thereafter it was a matter of sheer chance as to the value, if any, of the ledge laid bare. The second method was to select a likely looking place and to start sinking a shaft. Again the prospector knew that the quartz bodies almost never stood perpendicular to the surface. They slanted or dipped anywhere from a few degrees to almost ninety degrees with the horizontal. Therefore, the prospector had but to dig deep enough and a ledge would be exposed. Again it might, or it might not contain gold in paying quantities.

The second method was known locally as "coyoting." It was first used on Coyote Flat, just north of Deer Creek in Nevada City. There the placer miners found that bedrock was almost paved with gold. It was so rich that each miner was allowed only one claim, and that could measure not more than 10 feet on each side; that is, it could contain no more than 100 square feet. The miner dug straight down to get at the gold. He threw the dirt out in a mound at the side of his shaft. When bedrock was reached he scraped the bedrock, then he removed the overburden on all sides of his original "well." Late in the summer of 1852, a visitor stood watching the feverish activities of more than five hundred men as they threw the dirt out of their shafts to get at the gold. He is supposed to have remarked that the whole scene reminded him of a bunch of coyotes digging in for the winter. The name stuck, and thereafter any prospector digging a shaft in anticipation of hitting pay dirt in any form was said to be "coyoting."

In the Grass Valley area the coyoting method was favored over the trenching. That was because once a gold bearing ledge was found the miner was in position to go to work immediately. All quartz mines are worked from the bottom up. That is, when a ledge is to be removed to recover the gold from the quartz, a tunnel (drift) is dug at some depth (in an operating mine, usually at 100 foot intervals). The overhead valuable material is then blasted loose (stoped) and allowed to fall into chutes and storage bins from where it can be hauled to the shaft and hoisted to the mill.

One of the hard facts of "lode" mining (quartz vein) is that most ledges decrease in value at a depth of only a very few feet from the surface. This is known as "surface enrichment." This phenomenon is not limited to California, it occurs world wide. Historically, the most famous case was in Mariposa County where General John Fremont owned several thousand acres on the barren, or at best, brush-covered hillsides. There many of the quartz formations of the lower Mother Lode came through to the surface and were washed clean by the rains of the winters. Some of these ledges carried gold in large amounts.

Fremont started a number of quartz mines only to find that at a depth of no more than one or two hundred feet no more gold was found. At those points the surface enrichment stopped. Fortunately, Fremont had the good sense to stop mining before he had lost all the money he had made from the enriched ore in sinking the shafts deeper, convinced that just a few feet more would lead to rich ore once more.

In the Grass Valley area, the surface enrichment was particularly great. Only a few of the famous ledges paid below the first hundred feet. To demonstrate this, one has only to walk over the hillsides of western Nevada County to show that this was true. Stand on almost any spot in this area. Draw an imaginary circle with a radius of 1,000 feet, and within that circle one or more (usually more) dumps of waste rock from a long abandoned mine will be found. The "lucky" miner who found the ledge began digging with dreams of great wealth dancing before his eyes, only to find that below a few hundred feet at most, his mine was useless.

In areas where random digging was not feasible, the classical method of prospecting was for the miner to pan the dirt from the slopes above stream channels which had shown rich placer gold. If there was a gold bearing ledge within a reasonable distance above the stream, the dirt would show some colors. These would become more frequent as the prospector got closer to the ledge where they had originated. If the traces of gold in the pan stopped suddenly, the miner would know that he had passed over the top of the ledge. He then returned to the spot where the last colors had been found and started to dig. With any luck he uncovered the quartz which had carried the gold. With even more luck, he would find that the source of the gold was not just an isolated pocket, but was a ledge which carried gold along its length and to a considerable depth.

A few mines were discovered by a miner having a dream and going out the next day to dig where the dream had shown the gold to be. If anything was found it was, of course, due to incredible luck. Or in some cases the miner had, by allowing his subconscious to work, associated something he had seen during his travels through the hills with something he had seen close by rich mines.

On the other hand, Pike Bell of Auburn is supposed to have made a good living for many years by pocket hunting. His method seemed to have been to sit on the front porch of his home above Auburn Ravine, and dream away the long sunny afternoons. At intervals he would shake off his lethargy. He would order his handy man to hitch up his team and buggy and drive him into the hills. He seemed to know exactly where he was going. At the designated spot, he would have the driver tie the horses to a convenient bush and would start to dig. Too many times to be coincidental, a pocket of rotted quartz would be found. It might contain only a few tiny colors, or often enough to keep Pike in semi-luxury; gold up to several hundred ounces would be dug out.

There is an awful amount of land lying within a radius of fifteen miles of Auburn. The chances of anyone going out, even before the country had been dug over by ten thousand prospectors and hitting gold even in the amount of a single dollar, would be about one in one hundred million! How Pike Bell made his living doing just that we do not know, and he never did tell!

CHAPTER XVII

THE EMPIRE MINE

Every mine has its own story. The truth about the discovery and development of many mines does much to prove the old adage that "truth is indeed stranger than fiction." In this and succeeding chapters the stories of several mines will be recorded.[1]

The main shaft, and the buildings needed for the operation of a deep quartz mine, are located almost exactly one mile southeast of the city of Grass Valley, California. The discovery of gold bearing quartz took place within days after McKnight made the original find on Gold Hill, which lies to the west across Wolf Creek. The first of the quartz was found by George D. Roberts. Because of the richness of his find, he immediately christened the rounded crest on which the find was made, "Ophir Hill."

In four short years the title to the claim Roberts staked passed through several hands, each transfer being at a higher price as the mine had already produced more than $900,000 in gold. In 1864, the mine was purchased by a group which incorporated. The new company was headed by William Bourn. By 1870, Bourn had succeeded in buying out the other stock holders and now owned the stock of the mine in his own name. Meanwhile, the mine had reached a depth of two hundred feet along the main ledge and was producing ore which averaged $60.00 per ton at the mill.[2]

---

[1]The stories about the mines are true. While they are not documented, any student of history can verify them.

[2] Free gold only. Sulfides were not saved.

The mine continued to operate without interruption but with decreasing yield per ton, until by 1878, it had reached a depth of 500 feet. At that point the mine was considered to have been worked out. Some of the world's best mining engineers, including John H. Hammond, advised that it be closed before more money was wasted in searching for new pay shoots.

At this point, William B. (Billie) Bourn, Jr., who had made some money mining independently on Osborn Hill, asked for and got full title to the mine as a birthday present from his father. He continued to work the main ledge on down through the so-called "barren zone." At a depth of 1,500 feet, and after five years of operating at a loss, he found highgrade rock once more. The profits began to roll in.

To Bourn's faith in the mine must be added much new knowledge of the science of mining that was developed just at the time it was needed most. In addition, new machines had been invented and introduced into the mines. This made deep rock mining much less a hand operation. It eliminated a slow and costly way of removing the hard rock from the depths of the earth.

The development of the high pressure water pump and the ability of industry to supply long lengths of steel cable were major contributions. However, the one invention which made it possible to open up large areas underground was the compressed air drill. To this must be added the refinements in the construction and the operations of the stamp mills for the crushing of lower grade ores at a profit.

The Empire Mine, under the ownership and direction of Billie Bourn and his successors, continued to operate at high profits until World War II. Thereafter, it was worked at a loss until it was finally closed in 1957. The continuous operation lasted 107 years.

The amount of net profits the mine produced is best illustrated by the annual reports of the mine. In 1918, during World War I, when costs were abnormally high due to the shortage of labor and materials, the mine shipped $1,500,000 in gold bullion to the Mint in San Francisco. The total cost of all operations during that year was $500,000! (At that time the income tax was 2%.)

Many more times the pay shoots pinched off, or the ledge was faulted and lost, but Bourn always had faith in the mine and the intestinal fortitude to risk the money needed to sink the shafts still deeper and to explore for still more rich chimneys of highgrade ore. These, fortunately were always found. One of the unique things about the Empire was that the deeper the paying ledges became, the richer they were, making it possible for the ores to be raised from two or more miles below the surface and still continue to produce a high profit.

In 1890, Bourn retired from the active management of the mine. He appointed his nephew, George Starr, to the post he had just vacated. George was educated as a mining engineer at the University of London. He was thus able to use and to improve all the new innovations which were coming so rapidly at that time.

Bourn retired to a fine townhouse on Nob Hill in San Francisco and to an estate in Woodside where he could meet and associate with the rich and the powerful of the Pacific Coast.[3]

---

[3]The Bourn Woodside estate, "FILOLI," was taken over in the late 1960's by "The American Heritage Foundation." It is now open to the public. The twelve acres of formal gardens and the large authentic Victorian residence will be maintained in perpetuity for the American people. [To those who watch the "soap opera," Dynasty, the ancestral home shown during the opening scene, is FILOLLI, Bourn's country home.]

He had led an active and exciting life. But retirement did not fit his temperament. With cash in hand and with credit in almost unlimited amounts available, he looked for something with a minimum amount of risk to occupy his time. He found it in the Spring Valley Water Company. San Francisco was a growing city. The wells on which it depended for fresh water were beginning to show an infiltration of salt water from the Bay and the ocean. The people were glad to have the millionaire from the mines bring them a dependable supply of drinking water.

Directly south of the city the famous San Andreas Fault, the one which destroyed San Francisco in 1906, had formed a 25 mile trench along the crest of the ridge between the Bay and the ocean. Some attempts had already been made to use this natural reservoir as a city water supply. Bourn had his engineers throw a dam across the northern end of the depression. He piped the water into the existing distribution system, and held the citizens of the city hostage for any amount he wished to charge.

Finally the citizens rebelled. They organized a water district and built the Hetch Hetchy Dam in the alpine valley north of Yosemite. Then they constructed a pipe line down out of the Sierra, across the San Joaquin Valley and through the Coast Range to the Spring Valley Reservoir which they purchased from Bourn.

How many millions Bourn made from the operation and the sale of the Spring Valley Water Company, is of course, something which the public was not allowed to know, but to those who know of the propensity of the Board of Supervisors of the city and county of San Francisco to be extremely generous with the tax payers' money, it must be assumed that Bourn's profits must have been at least a "King's ransom"!

We do know this much; with the profits he had

made from his San Francisco investment, and possibly with a few more millions from the mine, he visited the birthplace of his father in Ireland. There he purchased the famous Lakes of Killarney along with 1,200 acres of the surrounding countryside. Then with more of his millions, he converted all of the property into a great park with his Georgian manor house at its center.

The people of the Lake Region of Ireland were most disturbed by the "filthy rich Yankee" taking one of their most famous and most beautiful places for his own private use and enjoyment. Bourn had the last laugh however (that is if one can indeed laugh from the grave!). When his will was probated in 1932, it was discovered that he had left the entire Irish estate to his only heir, a daughter, to be held in trust until her death. Then it was to revert to the Irish Government as a National Park! So, the next time you visit the Irish countryside; pause and give a little "thank you" to the gold from the hills of California, and to the man who, in the end knew how to use (at least part of) that gold for a noble cause.

Before Bourn retired to San Francisco, he visited Ireland several times. While on one of his visits, he saw an Irish manor house which he admired very much. Upon his return to California, he commissioned his San Francisco architect to draw up the plans for an exact duplicate of the Irish house. Then he had the Grass Valley workmen build the house on a knoll which overlooked the mine buildings to the south, and over the valley of Wolf Creek and on to the crest of the snow-clad Sierra to the north and east.

In 1975, the Division of Parks and Recreation of the State of California purchased the Empire Mine. The purchase included all the remaining mine buildings, the old residence, the Empire Tennis Club courts and clubhouse, and more than

500 acres of the surrounding pine, cedar and fir covered hills. The "cottage" as it was called by the Bourns, or the "Mansion" as it is called by the people in Grass Valley, has now been repaired and restored. The interior is included in the tours conducted by the Park Rangers. It is well worth the trouble to inspect both the exterior and the interior of the rough mine-stone and brick structure. The formal gardens which were planted to complement the house with its broad lawns and its water falls and reflecting pool, have also been stripped of the vines which had overrun much of the grounds during the fifty or more years of neglect.

When the Bourns left the hills for the life in the big city, the new superintendent was given the great house for his residence. George Starr, the first superintendent, refused to live there. It was most comfortable during the summer months, the thick masonry walls shielding it from the heat, but during the winters it was impossible to heat. In true nineteenth century European tradition, it was heated only by fireplaces, one in each of the fourteen rooms! Starr built a "Julia Morgan" shingle-covered house close by the tennis club (unfortunately it burned with a total loss in 1924). When Fred Nobs took over the management of the mine in 1918, he too refused to live in the "Mansion." He built a "cottage" on Empire Street for his family.

George Starr was the first of the trained mining engineers to continue and to improve upon the sound mining practices introduced by Bourn. Soon after he assumed the management of the Empire Mine, The North Star Mine on Lafayette Hill, one mile west of the Empire, followed Bourn's example and engaged a man with professional training and experience to direct its operations.

Arthur B. Foote, an engineer with wide experience in directing mines in the western

United States and in northern Mexico, brought his knowledge and experience to Grass Valley.[4]

Together, Starr and Foote made Grass Valley the Mecca of all serious students of mining who were also interested in the practical use of the new techniques and the new machines then becoming available to the hard rock miners.

Herbert Hoover, America's most famous mining engineer, used his summer vacations during his student days at Stanford University to gain practical experience in the Grass Valley mines. He worked first for Bourn in the Empire and later for Foote in the North Star. Upon graduation his first engineering position was at the Sneath-Clay Mine on Gold Flat near Nevada City. On the strength of the glowing recommendations written by Bourn, Star, and Foote, he was able to get the appointment as chief engineer for the London Mining Syndicate and later to become the mining consultant for the Chinese Government.

Hoover added a new dimension to his profession. He refused to work for anyone, private mine developer or government, unless he got "a piece of the action." That is, he was to get a percentage of the net profits of any mine he recommended be developed and over which he held the right to direct its operations. He was highly successful. So much so, that he was able to

---

[4]Arthur B. Foote was the real character in Wallace Stegner's Pulitzer Prize winning novel "ANGLE OF REPOSE." The novel was fiction, so Stegner changed the name to Ward. However the basic facts are true. Foote built the "Julia Morgan" shingle covered house on Lafayette Hill where the fictional grandson of Ward (Foote) was supposed to have been living when he wrote the history of his grandparents. Arthur D. Foote, the real son of Arthur B., had only three daughters; no sons.

retire while still a relatively young man to assume the management of the distribution of food to the displaced people, at the outbreak of the War in Europe in 1914. The rest is history.

After leaving the Presidency, Hoover "played" with his first love, mining. He returned to Grass Valley repeatedly to renew old acquaintances and to study the new mining techniques which were being developed. While there he was always the house guest of Fred Nobs. Nobs and Hoover were classmates and close personal friends during their undergraduate days at Stanford. When Starr began to think about retirement, he looked for a successor. Immediately he thought of the young mining engineer who had worked for him years before. Surely, Hoover must be tired of running around all over the world, living under the most primative conditions, and being denied the amenities of civilization. He had made plenty of money. Now he should jump at the chance to come live among the pines and the firs on the slopes of the Sierra. There, the responsibilities would not tax his energies and he would be within easy travel distance to the cultural centers of the Pacific Coast.[5]

Hoover thanked him for the offer, but said, "Get Fred Nobs." Nobs was hired and managed the mine during the take-over by the Newmont Mining Corporation in 1928. Thereafter, Nobs managed the consolidated properties which consisted of the Empire mine, the North Star Mine, the Pennsylvania Mine, and all the other small mines along the Empire Ledge System which extended from the crest of Osborn Hill to most of the ledges under the city of Grass Valley.

Nobs served during the prosperous years of the

---

[5]The author had the pleasure of meeting Mr. Hoover during these visits. He was also on the Scott River during the Scott Bar incident.

Great Depression. After World War II, he was responsible for many innovations to reduce the amount of labor used and for the more efficient use of modern machinery in the mill to increase the tonnage processed. The stamps were eliminated, except for some preliminary crushing, and the more efficient ball mill introduced. The cyanide plant was closed and the sulfide concentrates saved by the "oil flotation process," then shipped to a smeltery. And under ground, most of the "days pay" workers were eliminated in favor of the "Tribit" system, the workers furnished the labor and the explosives, while the Mine furnished everything else. Then when the checks were received from the Mint, the proceeds were divided equally between the workers and the Mine.

Even with the new methods of mining, the mines continued to lose money. The Government refused to allow the gold producers to sell on the world market (about $50.00 per ounce at the time). The Miners Union demanded that the miners receive the same pay as did the copper and lead miners in the Great Basin States where the prices of those metals were subsidized by the Government. The Unions would not believe that the mines would close if they did not relent in their demands. They would not, so the mines ceased to operate.

Just before the decision was actually made to close the gold mines, Nobs retired. Johnny Mann, a mining engineer and a Vice President of the Newmont Corporation, was sent back to Grass Valley where he had begun his career, to perform the unpleasant duty of issuing the orders to "pull the pumps."

Mann had all the machinery brought to the surface. It was put on display in the mine yard. Mining papers the world over carried advertisements to the effect that the machinery of the famous Empire Mine would be sold to the highest bidder. Thus, after 107 years of continuous operation, the mine was no more.[6]

The bronze plaque on the gate post at the entrance to the Empire Mine's grounds states that during the 107 years of mining, $120,000,000 in gold was taken from underground and milled at this location. Actually the total amount was probably closer to $200 million. Beginning in 1929, the underground workings of the Empire and the North Star Mines were connected underground. A part, at least, of the rich ore from the North Star was hauled to the Empire shaft to be hoisted and milled. Later, all milling was halted at the North Star. All the ore was then hauled by truck to the Empire mill.

In addition, there is just no way of knowing how much highgrade ore was smuggled out of the rich stopes, and how much was taken from the mill in the form of amalgam. Some have suggested that the total must have approached 10% of the free gold in the quartz. As about one-half of the gold taken from the mine was in the form of sulfides, the total take must have been somewhere between 5 and 10 million dollars. Most men familiar with the practice of highgrading are convinced that the last figure is much too conservative. The activities of the highgraders will be discussed more fully in later chapters.

---

[6] Some idea of the size of the operations at the Empire may be realized when it is known that the used machinery from underground, some of it having been used for more than forty years, brought more than one and one-half million dollars! The mill, the compressors and the hoisting machinery had already been sold at negotiated sales.

CHAPTER XVIII

THE COUSIN JACKS

When McKnight found the gold bearing quartz on Gold Hill, he began mining by digging down along the top of the ledge. He threw out the dirt on either side and tossed the quartz into a pile. Fortunately, at this spot the quartz vein broke away from the "country rock" on either side with no resistance. But as is almost always the case, McKnight's ledge did not go straight down into the earth. It sloped (dipped) at an angle of about 30 degrees with the horizontal. It can be seen, therefore, that after only a few feet, the soil on the upper side of the ledge would become too deep to be removed easily. The material under the ledge could be removed only by undercutting the ledge.

The early miners partially solved this problem by digging a four by six foot shaft so that the ledge itself acted as the hanging wall (ceiling) of the shaft. The quartz was then picked loose and saved. This process was continued until the rock became too hard to be mined by pick and shovel alone, usually at a depth of from ten to thirty feet. At that point a tunnel (drift) was dug, again along the foot wall of the ledge. At some convenient distance from the shaft, the ledge was removed from the overhead dirt and rock, again from below the hanging wall. This exposed more of the ledge so that it could be broken down later, hoisted to the surface, and taken to the mill. This process was known as "stoping." The area from which the rock and the ore had been removed was called a "stope."

The earliest miners on Gold Hill were most

fortunate. They were able to sink their shafts to a depth of thirty feet or more before they encountered rock so hard that they were forced to discontinue until methods of hard rock mining were introduced. Even so, the quartz was so rich that they succeeded in recovering more than 30 million dollars within the first two years, and all that wealth came from the surface of not more than 250 acres on Gold Hill, and from a few other very rich veins that outcropped in the immediate vicinity.

At almost the same time that McKnight made his discovery, Fremont and his Mexican miners found gold in the outcroppings of quartz on or near the Mother Lode Ledge in Mariposa County. (The southern mines claim that Fremont's discovery was made sometime before October, 1850. No certain proof exists.) In any case, the barren hillsides to the south and west of Yosemite Valley had allowed the rains of winter to remove the soil, and weathered rock, so that only the hard, unaltered quartz was found near the surface. This Fremont's miners were not equipped to handle.[1]

---

[1]Freemont had a great stroke of luck. Early in 1848, he was ordered west once more by the Army. He chose the southern route via Santa Fe to San Diego. Enroute he encountered a group of Mexicans who heard of the gold strike on the American River. They had mined placer gold in Sonora, Mexico, all their lives. They had the tools and the know-how so badly needed in California. Fremont remembered the 30,000 acres of worthless land he was tricked into buying three years beforc. Could there be gold on it? He guided the Mexicans to his land and allowed them to mine on a royalty basis. They found gold in the streams, and Fremont became a very wealthy man. At one time he was reported to have more than 500 pounds of dust and nuggets stored in bags under his bed in Monterey!

Fremont sailed immediately for London via Panama. London at that time was the center of mine financing and the manufacturing of mining machinery. Fremont purchased a stamp mill and the other tools used in the hard rock mines of Europe and Africa. Fortunately, the salesman who sold him the equipment asked if he had the men who would know how to operate the tools. He was forced to admit that he did not. He was advised to journey to Cornwall and engage some of the Cornish tin miners. They were reputed to be the best hard rock miners in the world.

Fremont was able to get a dozen of the more adventurous miners to accompany his newly purchased mining machinery around The Horn on a Yankee Clipper ship. At Mariposa, they set up the mill, showed the Mexican miners how to use the tools Fremont had brought from London, and proceeded to make Fremont a still richer man.

But all was not as it first seemed. The quartz ledges on Fremont's property carried gold to only a very shallow depth. After a few feet, the surface enrichment tapered off and the mine stopped paying.

Fremont offered to pay the return fare back to England for his Cornish miners, but they had seen enough of California to know that they wanted no more of the fogs and rains of the southern coast of Britian. In addition, they had heard of the quartz mines which were being opened on a small tributary of the Bear River, 150 miles to the north. They knew that they would be needed.

The relatively slight slopes of the hills near Grass Valley did not allow the top soil to be washed away as it had been near Mariposa. In addition, the heavy cover of pine, cedar, and fir trees carpeted the ground with a foot or more of decaying needles and branches. As the rain of winter seeped down through this decaying material, it picked up various acids which hurried the weathering of the rocks below. Consequently, there was a booming mining industry already in progress when the Cornish miners arrived.

But the Cornishmen did arrive at the right time. Most of the mines had just reached the hard rock. They were about to close, or were struggling to overcome the problem of sinking the shafts into the unweathered rock below. The former tin miners were immediately put to work showing the northern miners how to drill holes into the hard rock, fill them with blasting powder, and fire them in sequence to break out the solid material.

They also showed the local carpenters and blacksmiths how to build copies of the famous Cornish pumps which kept the deep tin mines in Cornwall free of water. This was most timely as at the point where the porous altered rock near the surface gave way to the solid rock, the mine shafts often turned into wells!

The Cornish miners were so successful that almost at once they were asked if they knew of any more good miners who might come to Grass Valley to work in the mines? The answer inevitably would be, "Well, me Cousin Jack back in Cornwall wouldst like to come 'or 'ere. I couldst get 'im to come wouldst ye pay 'is way, and guarantee 'im a job." So was born the name "Cousin Jack" for the Cornish miners and for all their descendants. What could be more natural than when their wives joined them in the new country, they immediately became "Cousin Jennies."

The history of the Cornish tin miners can be traced back to the time of the Phoenicians. Those rovers of the seas, exploring the southern shores of the British Isles, found the tin needed to alloy with the copper which they had in great abundance. This formed bronze, which for a time, they alone knew how to fabricate into tools and weapons. With their superiority over the other peoples of the "civilized" world, they controlled the seas and, therefore, the commerce of the known world.

The Phoenicians made slaves of the Britons and forced them to load their ships with the precious soft white metal. They carried it back to Asia Minor; where, among other things, they used a part of it to fashion the bronze fittings and decorations for King Solomon's Temple in Jerusalem.

A few centuries later, it was the need for the tin of Cornwall which led Caesar to interrupt his invasion of Gaul to cross the English Channel and conquer the wild tribesmen of southern England. He also must have the tin to make the bronze to arm his Legions with which to conquer the world. As all good generals must, he made sure that his rear was protected and that the routes to his sources of critical supplies were kept open.

With the advent of the "Iron Age," the importance of tin diminished, but never to the point where the Cornish miners were no longer needed. Thus, when the California mines needed their skills, they were available with the knowledge they had acquired during more than 3,000 years of following the ledges of cassiterite, the ore of tin, through the hard rock formations of Cornwall.

The Cousin Jacks were unique in a number of ways. They were slow to assimilate into the American way of life. Until mining came to a halt in the hills of California, one had but stand on the street of any town, listen to the conversation of the passers, and point out the people of Cornish ancestory. Most of them could speak good American when they wished. The teachers in the mountain schools had stood for no foolishness. The youngsters learned to know that there were such letters as "h" in the alphabet and when the teacher was apt to be listening, pronouns became "he" and "his," not 'e and 'is! But even after returning from college with degrees and documents to prove that they were doctors, lawyers, or even school teachers, they would, when talking to their own kind, revert back to their picturesque language.

Not too many years ago a Cousin Jenny, in a moment of frustration after she had accidently knocked a china figurine off the counter of a store and watched it break into a hundred pieces as it hit the floor, said, "E's bet wake, idn't 'e?" To translate, she had just remarked about the fragileness of the bit of china. "Bit weak, isn't he?"

One idiom which even the most strict teacher was never able to eliminate from the language, was the substitution of the preposition "to" for "at." For example, the ordinary American boy when embarking on an athletic trip might ask the coach, "Where are we going to eat AT, coach?" Incorrect English, of course, but the Cousin Jack boy, under the same circumstances would compound the error. He would say, "Where are we going to eat TO, coach?"

For two or three generations the Cousin Jacks remained quite clannish. Part of this was from necessity. No other ethnic group made up more than a slight percentage of the population except the Irish. Clashes were inevitable. The Cornish excelled in mining, the Irish in sports and in all other physical activities. All types of contests were scheduled. Mostly the Irish won without too much trouble. If they did, everything was fine. All was right with the world. If on the other hand, for some strange reason, the Cousin Jacks began to get the upper hand, the fighting nature of the Celts took over and a "free for all" developed.

To make matters even worse, the Irish were,

of course, mostly Roman Catholic, and very devout ones at that. The Cornish, almost without exception, belonged to the Methodist Church. Mostly they were subscribers to "The Menace," the violently anti-Catholic weekly. They firmly believed that the local branch of the Knights of Columbus stored arms and ammunition in the basement of the old St. Patrick's church, or when it was pointed out that the church had no basement, they countered with the information that the arms were hidden in the basement of Serra Hall. These were to be used when the Pope gave the word for the faithful Knights to take over the United States.

Until quite recently, there were few if any inter-marriages between the two groups. When in 1905, Mike McGuire married Bertha Colton, a sweetheart from the first grade at the Union Hill School, the gossips' tongues ran on and on. It must have been a "shotgun wedding." Only after six months had passed with no visible evidence of pregnancy and old Doc Jones let the remark drop that there had been no miscarriage, did the Cousin Jennies admit that it must have been love which had seduced the beautiful Methodist girl into joining the "Pope loving, statue-worshiping, Romans," and to send her children to a nun-taught school.

As late as 1928, firey crosses were burned on the hillsides of western Nevada County. Presumably by the Cousin Jacks, some of whom were known to belong to the anti-Catholic branch of the Ku Klux Klan, an organization then active in the southern and western states.

It was not by chance that there was a mistake in the wording of the plaque on the gold discovery monument at the top of Gold Hill. The monument was erected in 1929 by the local lodges of the Native Sons and Native Daughters of the Golden West. The memberships of both consisting almost exclusively of people with Cornish ancestry. They knew very well that it was George McKnight, and not George Knight who had found the gold in quartz, but damned if they were going to allow an Irishman of the Catholic faith to get the credit for that most important event![2]

The clannishness of the Cousin Jack boys made it sometimes unhealthy for the non-Cornish boys to come into town alone. Too often they were put upon and chased back to the country with black eyes and bloody noses. The tendency carried over into a bitter rivalry between the "poor" Cousin Jack boys of Grass Valley and the "rich" kids from Nevada City. Mostly, the families of the working men lived near the large mines in Grass Valley; on the other hand, the politicians, the attorneys, and the more influential businessmen lived in Nevada City. The big money of the County gravitated to the County Seat. They planned it that way!

To those who might question such a statement, a casual drive through the towns today will show that Nevada City still has a disproportionate number of Victorian homes of style and grace, clearly the residences of the wealthy. Grass Valley has the small homes of the miners.

The picturesque name of the hill crest midway between the two mountain towns of Grass Valley and Nevada City was acquired because of the semi-gang wars between the boys of the towns. Mostly, the

---

[2]The writer was on the spot at the time the monument was set in place. He read with astonishment the words on the plaque, but being a local school teacher, he kept his mouth shut. The members of the School Board were all Cousin Jacks. Later, when he researched the history of the area, he found that there was no doubt that McKnight was the correct name.

elders did not interfere with the "kid" fights. A few scuffles were a part of growing up. But when the miniature warfare seemed to be getting out of hand, the wiser heads of each community held a conference to iron out the differences. What was more natural than that they meet in the saloon at the top of the grade between Glenbrook and Gold Flat? The towns gathered to talk it over. Hence, the naming of the hill top, "Town Talk."

The Cornish miners brought with them not only their skills as underground workers, but their great love of music as well. When four or more Cousin Jacks gathered together, they began to sing. First as a quartette, and then as a male chorus. Until 1967, the Cornish Choir of Grass Valley gathered weekly to rehearse the "old songs," those composed hundreds of years ago in Cornwall and handed down through the generations from father to son. Towards Christmas time the activities increased, preparing for the great events of the season. Sad to say, in 1967, Harold George, the long time leader of the modern choir, found it impossible to continue. After the last Christmas Eve "sing" at the corner of Mill and Main Streets and the final appearances in the churches, the famous organization was disbanded.

The choir was unique in that it was composed of boy singers along with the mature male voices. At the age of twelve, the Cornish boys were introduced. They sang the alto parts until their voices changed. It was not unusual for three generations to be singing together. In fact, almost always several family groups made up the personnel of the choir.

During the mid-thirties, the choir broadcast Christmas Carols over the National Radio Hookup from the 2,000 foot level of the Idaho-Maryland Mine in Grass Valley. World War II stopped that unique program, but the choir did continue to serenade the townspeople, and they did continue to travel to the San Francisco Bay cities where they performed in the churches for the benefit of the misguided Nevada County folks who had been mislead into following the elusive trail of fortune and deserted the quiet life of the mountains for the rush and the smog of the cities.[3]

The Cornish miners were artists with both the double and the single jack. These were the heavy hammers used to strike the drills as the necessary holes were bored into the hard rocks of the mines. The miners knew that the hard rock could not be broken or drilled by a sharp pointed instrument. They also knew that the individual crystals of which the rock was composed were brittle and could be "exploded" by sharp repeated blows. So they developed a steel rod, one end of which was sharpened somewhat like a carpenter's chisel, except that the cutting edge was formed across the center of the rod rather than along one edge.

The method of operation was to hold the rod with one hand, pressing it against the rock to be drilled, and striking the other end with a four to five pound hammer (single jack) held in the other hand. After each blow the drill was rotated about twenty degrees so that the cutting edge now rested on new, unbroken surfaces.

Using this method a good man could drill a hole one inch in diameter into solid granite at the rate of eight to ten inches per hour. A

---

[3]All too often the Cornish folk music is classed as, or mistaken for, the songs of the Welsh coal miners. This error is made by some musicologists. The music is indeed similar, but there is a distinct difference. Fortunately, since the retirement of Mr. George, the Music Department of the Nevada Union High School has consigned the ancient music and the words to the printed page and carried on the tradition of the "Christmas Sing."

minimum of sixteen holes were needed in the face of each drift before the rock could be blasted out. It needs only a small knowledge of simple arithmetic to show that a man would need about five full days to do the necessary drilling for each four feet of tunnel.

The "double jack" was a ten to twelve pound sledge hammer. It was equipped with a thirty-six inch handle and was swung like an axe in a loop over the head to get the full force of the arc. It was held with both hands; so an additional man was required to hold the steel and to rotate it after each blow.

On a modern construction job, a double jack is sometimes used to break holes in concrete when the job is too small to get a jack hammer (pneumatic drill) on the job. Because the modern double jack man is not skilled enough to allow the steel man to hold the steel drill with his hands, a pair of tongs is provided. Without the tongs, one miss and the hands and the wrists of the holder would be crushed. The old time Cousin Jack miner would have scoffed at such "panty waist" methods. He had absolute confidence in the hammer man. He knew that he would never miss the blunt end of the drill and that, come his turn to swing the hammer, he damned well better not miss either. If he did, he would never be allowed to go underground again. He concentrated on the task at hand.

The steel holder had it too soft for the hard-driving mine bosses, even though he was taking his turn at swinging the hammer; so a second hammer man was added and the two hammers struck with a rhythmic beat. Later, still another hammer man was added if and when the working space was available, as in a large shaft. With three hammers going, the size of the steel drill was increased to one and one-half inches in diameter. This was to allow more blasting powder to be tamped into the drill holes. Consequently, much deeper holes could be used. A good miner knew how to space the drill holes, how deep to drill, how much explosive to put in each hole, and in what sequence to detonate each charge to get the best results.

During the Depression, a large number of new mines were started, and many old mines were reopened. Many of them failed due to the lack of experienced miners. Thousands of men were available, but being inexperienced, they were often worse than useless. The offer of higher wages failed to entice many of the real miners away from their stable jobs at mines with long histories of production and of giving employment during good times and bad.

In 1938, Mike McGuire was asked to take charge of the shaft men at the Shingle Springs Mine in Placer County. He inspected the mine, and although offered $300.00 per month—a fabulous wage at that time—he refused to work for wages alone. (Mike, of course, was Irish, but he had learned his trade under the best Cousin Jack "Captains" in the Empire Mine. Captain, being the name given to the shift boss by the Cousin Jacks.)

The mine was sinking a double compartment shaft (six by sixteen feet) through an altered quartzite which was so hard that the inexperienced miners were able to deepen the shaft only about two feet per day. This was prohibitively costly. Something had to be done or the mine would be forced to close. Mike watched the shaft men at work for a full day. Then he propositioned the management. He agreed to take complete charge of the men working the shaft. The mine was to give him full authority and supply all the materials needed.

He made only two demands. He must be allowed to hire one Finnish miner for each shift. These were men from the Michigan copper mines who had lost their jobs due to the closing of all the mines in the Mid-west. He had used them on a job he had just completed and knew them to be almost as skilled as the Cousin Jacks, and of more importance, they were available. The second demand was that he was to be paid no wage. Instead, he was to receive the difference between the cost of the powder (dynamite) that the inexperienced miners had used and what he would use to get the same footage. Finally he guaranteed that he would sink the shaft at the rate of six feet per working day. He also was to receive a bonus of $10.00 per foot for every foot he averaged over the guaranteed amount.

He personally directed the spot on which each of the holes were to be drilled to a depth of eight feet using bits with a diameter of two and one-half inches, then he sent up for the proper amount of powder to be lowered.

The men who had been running the job before he had taken over sat back prepared to have a good belly laugh at the expense of the "wise guy" who thought he was going to show them up. They were positive that, although he had mined all his life, he had never worked in any ground as hard as this, and from their experience they knew that the small charges he tamped into the drill holes would do no more than "rifle" the tamping material out of the holes in the quartzite.

Mike had the last laugh. He lit the fuses so that the charges would go off in the proper sequences, cutting out the center of the shaft to a depth of about three feet. Then he fired the deep corner holes, breaking the hard rock away from the walls in just the planned amount. After 24 rounds had been fired, eight feet of broken rock occupied the bottom of the shaft, not a piece of which weighed more than ten pounds. The "swing" shift of muckers soon loaded the broken material into buckets and had it hoisted to the surface. The rock was so hard that no "graveyard" shift needed be used for timbering.

The job was done in six months. Mike's check averaged $1,000 per month. This was more than twice as much as the superintendent was getting, but the owners of the mine admitted that it was the best money they had ever spent.

Hard rock mining is an art as well as a science. The old saying around the training quarters of the fighters is, "The bigger they are the harder they fall!" The good miner knows that, "the harder the rock, the better it breaks!"

WASTE ROCK SKIP, EMPINE MINE,
NOW DISPLAYED AT MINING MUSEUM
NORTH STAR POWER HOUSE

CHAPTER XIX

THE STAMP MILLS

As soon as hard rock mining began, the problem of effective crushing of the quartz containing the gold became a determining factor in whether or not a mine could operate at a profit.

During the time that only soft, well weathered rock from near the surface was being mined, it could be crushed by hand using a hammer on the surface of a large flat rock or by the use of an iron rod in an iron kettle. This led to some unknown blacksmith making an iron mortar and pestle, the idea coming from the use of such a device by a druggist to grind his chemicals and his herbs. Soon, the blacksmith made the device so large that the miners were forced to use two hands to lift the pestle. A two quart steel mortar is still standard equipment for hard rock prospecting.

From the date McKnight made his discovery until a few days before Christmas, the partners who had found gold in the rock they had saved to build their fireplace, pounded out $30,000. They did this by taking a ten foot pine pole, six inches in diameter, and having a blacksmith "shoe" it by fitting an iron cap over its larger end. They tied the pole to a springy oak limb which grew conveniently out over a large flat rock close by Wolf Creek where there was a constant water supply. Then they hollowed out a depression in the top of the rock in the manner of an Indian meal mortar. The device was operated by putting a few pounds of ore in the hollowed-out space and pulling down the iron-shod pole by throwing their weight on it. The spring of the oak limb pulled the pole back up, ready for the next stroke of the stamp.

The group soon made an improvement in their crude stamp mill. They cut the iron shod pole loose from the oak limb and tied it to the end of another pole by the use of rawhide thongs. This pole they then balanced on the top of a post. The second pole now acted as a lever, allowing one man to lift the iron-capped stamp with ease by pulling down on the long end of the lever. He then let go. The stamp pole fell back into the crushing sump. This innovation allowed the use of much larger stamp poles shoed with heavier iron caps. Naturally, the heavier the stamp, the greater was its crushing effect.

This was dry grinding. As soon as the ore being ground was judged to be sufficiently fine, it was scraped from the mortar, carried to the water and panned. In all, the partners recovered about 2,000 ounces of gold!

When they returned after living the rest of the winter in San Francisco in the most luxurious accommodations available at that time, they made an additional improvement on their "stamp mill." They dug a ditch along the bank of the creek to carry a constant stream of water to their stamp. Then they put a small sluice box below the large rock, and eliminated most of the work of panning to separate the gold from the sand.

The first mechanical device for crushing the ore, that did not require man power to operate, was the arrastre. Again, as with the panning bowl, it was the despised Mexican who showed the way. Who invented the arrastre (Americanized to arasta) is not known. It probably was brought into Mexico by Cortez. Certainly, for a very long time it had been used by the Mexican silver miners. They were quick to adapt it to the gold mines of California when the need for grinding very hard ores developed.

The arrastre is merely a large flat, very hard rock, fastened by a chain to a long beam or pole. The chain was fastened to the pole close to one end. That end was then fastened to a short post which was free to turn in a circular motion. To the long end of the pole, was hitched a mule or some other animal motive power, either a horse or an oxen. (In the early days in Mexico, the Indian slaves were hitched to the pole and spent their entire lives trudging around in the endless circle.)

The arrastre rock was dragged over a pavement of closely fitted, flat- topped cobble stones. The chain affixed to the front of the arrastre stone was made short enough to cause it to lift the front edge two or three inches above the pavement. This allowed the ore to be pushed under to be crushed by the weight and the movement as the arrastre stone was pulled forward. Again, the abrasive action of the heavy rock above ground the ore to a fine sand. It was then swept up and panned.

A mule and a man could crush, grind, and pan about 200 pounds of hard rock in a day by this method. To make it pay, only fairly rich rock could be mined and treated. The highly enriched rock often ran well over $1,000 per ton. Hence, a man and his mule could make a hundred dollars a day, if he had good ore.

The Mexican arrastre was a great help, but it did not solve the problem of the Yankee miner who wanted to get rich instantly. For the Mexican who was used to making just a bare living by grinding a little copper or silver ore, the dragging rock was good enough. The Americans had other ideas. They tried to speed up the grinding by making the grinding rock larger and heavier, and by having it pulled by two or more animals. They built them in "batteries," that is several arrastres built side by side, but they found that the Mexican method was not adaptable to large scale operations; so they returned to the idea of the stamp mill.

A Yankee blacksmith who had been watching the hand operated stamp, set to work in his shop in Boston Ravine. He took the same basic idea of using the iron capped pole for its crushing effect. Instead of raising it by hand and letting it fall on the ore, he fashioned an iron collar around the pole about a foot from its upper end. Next, he rigged up a water wheel on Wolf Creek. To the outer rim of the wheel, he fixed a projection so that, as the wheel revolved, it hit the iron collar from below and lifted it about 13 inches. Then, as the wheel continued its circular motion, the projection slipped past the collar and the stamp was allowed to fall.

The blacksmith was not satisfied with the Indian mortar rock under his stamp; so he fashioned an iron mortar to hold the ore and to catch the force of the blow. Then he made yet another improvement. He removed one side of the mortar box and fitted the opening with a metal screen. Now he borrowed the running water idea, and led water into the mortar box. The continuous sloshing of the water caused by the falling of the stamp allowed the fine sand and the small bits of gold to work through the screen and into a sluice box. The grinding operation was now at least semi-automatic.

A short time later the miners learned to put a small amount of mercury in the bottom of the battery box. This caught most of the gold before it escaped into the sluice box. Periodically, the stamp was stopped and the gold amalgam taken from under the stamp. Still later they eliminated the sluice box and the panning by placing a copper-covered table below the stamp. The "pulp" coming through the screen was allowed to run over the top of this table. The fine gold, being still heavier than the water and the sand, was trapped by the mercury which had been added to the surface of the copper. A gold amalgam was formed which was easy to recover.

Other improvements were in the process of being made when Fremont returned from his machinery buying trip to London. Among other things he imported a fully operational stamp mill.[1] Copies were soon made and duplicate machines were manufactured by the foundrys which sprang up in San Francisco and in the quartz mining centers. (To Hell with the patents. Who was going to prosecute someone in California, 10,000 miles away?)

---

[1]A bronze plaque, on the brick wall of the old Wells Fargo building in Mariposa, states that John Fremont found gold in quartz on his land in the fall of 1850. He journeyed to San Francisco and purchased the first stamp mill to be installed in California. He erected it here and ground the first gold- bearing quartz mined in California.

The controversy over who should have the honor of having discovered gold in quartz, McKnight or Fremont's Mexican miners, will probably never be settled.

It is clear, however, that the claim of Fremont having started to grind ore by the use of a stamp mill is highly questionable. Fremont, himself, made no claim of beginning to mine quartz ore until late in the fall of 1850. By what miracle could he have gone to San Francisco, purchased a stamp mill, and had it hauled the 150 miles over non-existent roads to Mariposa, and have it grinding ore in such short a time? And by what farther miracle would there just have happened to be a stamp mill in San Francisco for sale?

The closest place where such a machine could have been manufactured was in Manchester, England. A minimum of six months would have been required

The stamp mills which are still being used today in some small isolated mines are basically the same as the one brought from London by Fremont. There are five stamps, each weighing anywhere from 200 pounds to 2,000 pounds, operating in a single battery box. The stamp heads and the "anvils" on the bottom of the battery box are replaceable. This is a necessity because the sharp quartz crystals cut into and wear away the softer iron or steel which makes up the stamp and the anvil below.

The individual stamps are fitted with an iron collar near their upper ends. This allows a series of five cams, mounted on a single shaft, to raise and let drop the stamps in the proper sequence. To the end of the cam shaft is attached a pulley. Power from an electric motor, from a water wheel, or from a steam engine is transmitted over a belt. The size of the pulleys on the motor and on the mill are varied to give about twenty-two and one-half RPM. The cams are double acting. That is, there is a projection on either side of

---

[Footnote 1, continued] to bring it around Cape Horn and on up the coast to California. In any event, we have Fremont's own account of having sailed to England to find and buy the machinery needed to mine the hard quartz, and to extract the gold from the ore once it had been mined.

What undoubtedly happened was that the Mexican miners built an arrastra, and began extracting gold by mule power late in the fall of 1850. Unfortunatley the writer who drafted the wording on the plaque was not familiar with the difference between an arrastre and a stamp mill. Hence the error, to be read and believed by the townspeople, and by the 100,000 tourists who stop to read the plaque each year when enroute to the valley of the Yosemite.

the shaft. This allows the stamps to be raised and dropped at the rate of 45 times per minute.

The anvils are made slightly smaller than the space in the bottom of the battery box. This allows mercury to be used in the space below the top of the anvils. Larger chunks of gold, as they are broken loose from the quartz, drop into the mercury to be trapped for easy recovery. The fine gold, that which has a tendency to remain mixed with sand and water, is separated by running over the mercury-coated copper table top.

By 1916, the Empire Mine had 16 batteries, or 80 stamps in operation. Each stamp weighed 1,750 pounds. These were operated 24 hours a day, three hundred sixty-five days a year. Only one battery was ever closed down at one time for repair or for the weekly "clean-up."

The distant roar of the pounding stamps could be heard distinctly at the center of Grass Valley, two miles away. The townspeople became so accustomed to the noise that they paid no attention to it unless for some reason the rhythm was changed.

At one-fifteen on Christmas morning 1935, the switchboard of the town's telephone office suddenly lighted up as several hundred people tried to call the local police, the office of the Empire Mine, or just information. They were all demanding to know why the mill at the Empire had shut down?

It seemed that one of the workers in the mill had had too many "Tom and Jerries" before going to work on the midnight shift. As he began his tour of the stamps to be sure that they were working correctly, he found that the broken ore was not feeding down from the bin above into the chute of the automatic mill feeder. He took a crow-bar and climbed into the bin to loosen the lodged ore. Suddenly the whole mass of some 2,000 tons broke loose and came crashing down, pinning him against the lower wall of the bin. Had he been sober, he most probably would not have gone into the bin alone, or, in any case, he would have noticed the ore mass beginning to move. In that case, he could have made a quick escape up the ladder provided for that purpose.

As it was, when the mill boss missed him, he began a search expecting to find him "sleeping it off" in an isolated corner. Instead, he found one of the worker's hands sticking out above the ore. He immediately pulled the switches to close down the mill so that every available man could help dig out the "damned fool", hopefully still alive.

The minute the rumble of the mill stopped, the strange silence awakened the sleeping people just as completely as would an alarm clock. Most of them could not go back to sleep even after being told what had happened, until the mill began its usual pounding once more.

The man? Oh, he was crushed by the impact of the avalanche. He died almost instantly!

The stamp mill was made to order for the mill man who wanted to augment his meager wage by helping himself to some of God's riches. That was the reasoning of the mine workers who agreed with the Cousin Jack who freely admitted that the "Good Lord did not put all that beautiful yellow metal in the hills for the exclusive use of the rich mine owners." Consequently, it was no sin to spread the wealth around a little.

Before mechanical feeders were put into use, the mill man had to feed the stamps by hand. He could tell by the sound of the stamps hitting the anvil if and when more ore needed to be put into the battery box. When needed, the ore was shoveled out of a pile into the mill. Being human, the operator fed the correct amount of ore by listening. His eyes he reserved for the more important task of watching for the glint of gold in the ore as it dropped off the point of his shovel into the stamping chamber. By instinct, he grabbed at any piece containing gold. All too often he was forced to reach under the falling stamp rather than see it escape his clutching fingers.

If he had miscalculated the rhythm of the stamp, the hundreds of pounds of the stamp would shear off the fingers as neatly as would the wheels of a steam locomotive. Mostly, however, the mill man would beat the stamp. The speed of the mill allowed him almost a second and a quarter, plenty of time if he timed it right. As the mill man grew older, however, his reaction time slowed down and after long years of experience, he was bound to get a little careless. It is said that the older mill men could always be spotted by the missing ends of the fingers on his right hand; or on the left hand if his instinct was to reach from the "port side."

The invention of the automatic feeder saved many fingers. It also put a stop to that particular form of larceny.

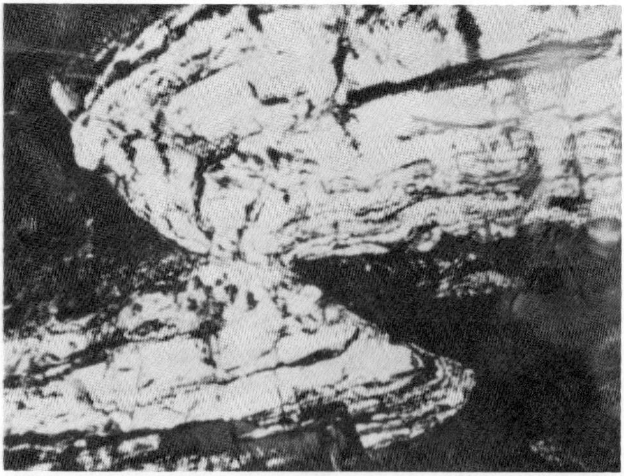

Brunswick Mine, Mother Lode, Quartz ledge showing reverse fold. — Photo Courtesy of Nevada County Historical Society

CHAPTER XX

HARD ROCK MINING

For some time following the discovery of gold in quartz, the lack of an efficient method of drilling holes in the hard rock in which to place the explosive charges was a major problem for the men who were trying to get the ores out of the ground and to the mills. It took "real men" to swing a single jack or a double jack for the ten hours which was the customary working day in the mines. In the granite rock of the mining country, even the most experienced and the hardest working man could drill no more than six inches of one-inch holes per hour.

The first mechanical assistance the miners had was the steam hammer. This machine operated on the same principle as the modern jack hammer or pneumatic drill. (The name "jack hammer" being a carry over from the single and the double jack which it replaced.)

The steam hammer when used to drive the drill into the rock was a great improvement over the strong arm method, but it is easy to see that the necessity of getting live steam to the working faces deep underground presented some serious difficulties. With wood or coal being the only possible source of fuel, generating steam underground in the working areas was, of course, impossible. With the boilers placed on the surface and the steam piped to the machines below, the great loss of heat, even in well insulated pipes, made the method very costly. In addition, working with live steam in the confined quarters of a drift or a shaft, made steam dangerous and impractical.

When someone came up with the bright idea of substituting compressed air for the live steam and produced the pneumatic drill as we know it today, the mining men of the world breathed a sigh of relief. Now air was compressed on the surface, using steam or water power, and piped to the men below. At that point, hard rock mining took a giant step forward.

In addition to furnishing the power needed to operate the drills, the mechanical loaders, and all the other machines used in the mines, the compressed air gave up the energy it carried, and provided fresh air to help keep healthful the air deep underground. Usually, of course, the spent air from the machines gave only a fraction of the ventilation needed by the miners. A regular ventilation system had to be provided to rid the mines of the tons of poisonous dynamite fumes generated during the removal of the waste rock and the ores from deep within the earth. These fumes, being heavier than air, would pocket in the deepest parts of the mines. Mechanical means of forcing them out was necessary.

Every reader of this material has seen a modern jack hammer in operation on our streets and highways. The machine consists of a piston which is pushed ahead by compressed air. This drives a drill or bit against the rock or the pavement to be broken. Valves are so arranged and activated that at the bottom of the power stroke, more compressed air is introduced to push against the lower side of the piston to return it to its original position, preparatory to delivering the next stroke. A heavy rubber hose leads from a compressor to the jack hammer. The operator holds the hammer by a double handle and gets a good shaking up as the bit or drill bounces off the hard surface being broken. Usually, the drill operates at about 500 strokes per minute. It delivers its blows with a force of at least 100 pounds per square inch, or 10 times the force a man can deliver with a single jack, and at 10 times the number of strokes per minute.

In the mines it is obvious that the common jack hammer could be used only in the sinking of a shaft where the drill holes are directed downward. Where the holes are to be drilled horizontally as in drifting, or verticlly as in stoping, a mechanical means of holding the device must be provided. This is done by clamping the machine to a bar which is wedged between the hanging and the foot walls in the stopes, and between the roof and the floor in the drifts.

When the hammer is clamped to the bars, the drills can no longer be pushed ahead by hand as when an ordinary jack hammer is used. To overcome this difficulty, a screw device is built into both the "drifter" and the "stoper." The miner operates the screw by hand, keeping the correct pressure on the drill as it strikes the rock. A certain amount of skill is needed to keep the machine operating at peak efficiency. The skilled miner knows by sound and by feel when the proper amount of pressure is being applied.

The screw device will allow the bit to advance about 18 inches. Once that depth has been reached, the drill must be removed and a longer one substituted. Then the drilling continues until another 18 inches has been drilled. The process is repeated until the hole has reached the desired depth.

The miner is provided with drill steel in various lengths, each succeeding length being two feet longer than the next shorter one. Usually the "starter" is from 18 to 24 inches in length. The next, 40 to 48 inches, and if the holes are to be drilled to a depth of 6 feet, the last steel will be about 72 inches in length.

Instead of swinging the single jack or the double jack, the miner now fits the proper steel drills into the machine and keeps the pressure on the steel by screwing the bit ahead as it bites into the rock.

The bits have to be re-sharpened after being used only a few times; so tons of steel are constantly being brought to the surface from underground, heated in a blacksmith's forge, and hammered into their original shapes to give the proper cutting edge once more. Then while still hot, the bit is plunged into a tub of water to restore its "temper." The steel is then sent underground where it goes through the whole cycle once again.

This constant grinding away of the steel by the hard rock below and the re-sharpening reduced the length of the steel bits. Eventually the four foot steel became the two foot steel, etc. Finally, the shortest bit is worn down to about one foot. Then it is of no more use, even as a "starter bit." It is discarded.

As time went on, several improvements were made in the design and the operation of the drill steel. First the drill rods were made with an one-eighth inch hole running through their centers. This was connected through the drilling machine with a high pressure water hose. When the machine is turned on, a jet of water is forced through the center of the bit and against the face of the rock being drilled. This stream of water does two things. First it cools the bit so that more pressure can be put on to make it cut more rapidly, and second, it keeps down the rock dust from the cutting of the rock.

Most of the California gold mines were so called "wet mines." That is, there was naturally enough water in the rock to keep most of the ground-up rock from being blown into the working areas as a fine dust. However, there were always some dry spots which, without the dampening effect of the hollow drills, would tend to keep the machine operators working in a cloud of finely powdered rock.

"Miner's consumption" was a very common disease in all the mining camps before the invention of the hollow drill. It was thought to be lobar tuberculosis, brought on by the breathing of the raw damp air underground, and by the lack of exposure to sunlight by the men who worked underground during the day-light hours. Modern medicine has shown that the disease was not the result of either of these. Instead, the "consumptive" miners are now known to have been suffering from "Silicosis." This condition is brought on by a concentration of powdered rock dust collecting in the miner's lungs. This rock, or silicon dust, is composed of millions of particles of very fine silicon crystals. These scratch the delicate tissues of the lung membranes. Because of this scratching effect, the tubercle bacilli, the cause of the disease, have easy access to the unprotected surfaces. They quickly take hold and grow.

Modern health laws do not allow "dry drilling" underground. Even on the surface the jack hammer operator must wear a mask if much dust is being generated. Modern workman's compensation laws also makes the assumption that if a miner becomes ill from silicosis, it is the fault of the mine and compensation must be paid. Most mines are now super-cautious to see that no dust develops underground.

Modern metallurgy has also developed drill steel which is so sophisticated that the local blacksmiths cannot sharpen it. Instead, the cutting heads of the drills are detachable, and when they eventually become too dulled to cut through the hard rock, they are detached and sent back to the factory to be refaced. Since the drill shafts are no longer being shortened by constant wear, only one set of two, four, six, and eight foot lengths of drill stems are needed for each drilling machine.

Eventually mechanical loaders were developed to lessen the back breaking work of mucking. For the most part, however, these machines were not of much help in the California gold mines. From the beginning, the deep quartz mines had operated with a minimum of waste rock and ore handling. Except in sinking a shaft, or in driving a drift ahead, the rock was almost always blasted from working faces in such cramped quarters that mechanical aids could not operate. In addition, except in the shafts and in the drifts, all mining operations were carried on from the bottom up. That allowed any rock broken loose by blasting to fall directly into chutes and into bins. Thus the miners let gravity do the hard work.

From the bins, the rock was allowed to fall directly into cars to be pushed by hand, hauled by mules, or pushed by electric locomotives to the main shaft of the mine. There it was again allowed to fall into larger bins from where it could be automatically loaded into buckets or skips to be hoisted to the surface. There it was either spilled over the face of the waste dump, or it was sent to the mill where it again was carried from process to process by gravity. (The photographs of every mill building, ancient or modern, shows that the building sits on a slope of about 30 degrees.)

Eventually safe electrical wiring for underground use was developed. This led to the replacement of the man and mule power previously used in moving the ore cars. Since it was mostly the mule power which the electric engines replaced, the new machines were immediately dubbed "electric mules" by the miners.

It was not until the shortage of manpower and the rise in miner's wages during World War I, that any attempt was made to do away with the "one man, one car" mode of operation. At that point it was found practical to purchase half-grown mule colts and lower them into the mines while they were still small enough to fit into an ore skip or to be trussed into a canvas bag and lowered directly down the vertical shafts.

Once in the mine, they were taken to already mined out areas which had been made into paddocks and stables. There they were allowed to mature, broken to harness, and put to work hauling the cars through the dark and damp drifts.

After a relatively short time, the mules became quite blind as the only light they ever saw was the dim flicker of the miner's carbide head lamps. They remained underground for their entire lives. They were brought to the surface only when death made it possible to truss the lifeless body into a shape which would fit into an ore skip. Then it was raised to the surface and buried in a "mule cemetery" near the mine shaft opening.

Their keepers and drivers, the mule skinners, often became quite fond of their charges. Their greatest show of affection was to share their cut plug chewing tobacco with their four footed friends. To the mules this was candy. Unlike the miners, they did not chew the sweetened aromatic weed, they ate it!

The replacement of the mules by the electric engines led to the eventual disclosure of a choice highgrading story.

Old Jim Barnes was for years an underground mule skinner in one of the larger Grass Valley mines. When the last mule was replaced by modern equipment, Jim was probably the most upset man ever to lose his job in the mines. The mine manager offered him a job on the surface, but he turned down the offer, stating that: "If he could not work with his mules, he was not going to work for the mine at all." This all happened before Social Security came into being, and the Mine had no pension system. Some of Jim's friends were really worried as to how he was going to live now that his wages from the mine had stopped. Others, however, had no illusions as to Jim's ability to live without working. They knew that years before he had developed the best racket in all of California for getting highgrade out from underground and turned into cash at the local highgrade gold buyer's house.

Jim's modus operandi was to buy the highgrade gold ore directly from the stopers who made sure that the most beautiful pieces did not get into the ore cars. He paid them about one third its Mint value. He then broke it into small pieces and sent it to the surface mixed with a skip load of mule manure which was sent up periodically to keep the mine clean. Now, of course, Jim never went near the manure pile. He had an accomplice who grew pears near Chicago Park. The rancher bought enough of the fertilizer from the mine to be considered a regular customer. It therefore aroused no suspicion when, once every month or so, he managed to be at the manure pile just as a very special load was being dumped.

He hauled the fresh stable cleanings to his ranch, but instead of putting it around his pear trees, he piled it near his barn, moistened it well, and allowed it to decay for a month or two. When the straw was well rotted, Jim rode out to the orchard on a weekend. There he put the gold ladened manure through a small sluice box. Now the pear trees were given their treat of nitrogen and other organic compounds. The broken highgrade was scraped from back of the cleats in the sluice box and its gold contents recovered by grinding the ore still finer and by panning.

Over the following weekend, Jim would visit still another friend, one who operated a small gold mine in Plumas County. The miner mixed the Grass Valley gold with the proceeds from his monthly clean-up and sent it to the Mint along with his regular shipment. When the government check was received, Jim got his share.

In a modest way, Jim had become quite wealthy. He had lived very frugally. His share of the $10,000 a year that came out of the mine along with the mule manure was put in a Reno bank and, eventually, into government bonds.

In his crude way, Jim loved his mules, but he was not above doing a little autopsy work whenever one of his charges joined its mare mother and jackass father in equine heaven. Before he sewed the carcass back together, he inserted as much as a hundred pounds of crushed highgrade in the dead mule's empty stomach. This amount was not sufficient to arouse the suspicion of the muckers who loaded the dead mule on the up-skip and buried the carcass in the mule cemetery.

Jim watched the burial and carefully marked the location of the grave. Then after a few days, and during the dark of the moon, he did a little grave robbing. When the carcass was exposed, a few quick strokes with a knife "liberated" the sack of highgrade. Stitching up the incision was not necessary. The only precaution Jim need take was to see that no sign of the opening of the grave was left for unwanted eyes to see. Some leaves and pine needles scattered over the fresh dirt did the job quite nicely. The highgrade was taken directly to the Plumas County mine. The orchard storage process was not needed.

Jim was almost killed one night when some muckers showed up to do a little ghouling of their own. They arrived at exactly the time that Jim was making the incision to recover the sack of highgrade. It seemed that they had buried a mule of their own in the grave immediately next to Jim's mule. They too had pumped some highgrade into its stomach, and they now were about to recover their treasure. In the dim light they assumed that they had caught a thief in the act of stealing "their gold." They jumped into the opened grave and proceeded to give Jim an unmerciful beating before he could get through to them that he was not robbing them. He was digging up his own mule!

Shortly after Jim retired, the story of his source of wealth became common knowledge among the miners. For the rest of his life Jim was affectionately known as "Mule Droppings Jim." At least that was the term used when ladies were present. In the pool-halls and at the local bars, the more primative Anglo-Saxon term was used!

Early Air Operated Stoping Machine (Brunswick Mine) — Photo Courtesy of State Bureau of Mines

CHAPTER XXI

SULFIDES IN THE ORE

The gold recovery systems used by the California mines were paradoxical. The mines were constantly looking for ways to cut a cent per ton off their costs. They were among the first of the large industrial institutions who regularly invested some of their income in research aimed at finding new ways to get the last cent of value from each ton of ore. Yet they stayed with the stamp mill as their principal means of grinding the ores long after that crude machine had become obsolete.

When one realizes that, year after year, the Empire Mine was bringing to the surface and processing more than 1,500 tons of ore a day, 365 days per year, the amount lost by inefficient use of machinery became significant. Ten cents per ton going out of the mill along with the tailings amounted to $150.00 per day! In the course of a year the amount lost became impressive.

In the first years of the mines, it was not cents, but dollars per ton that were being lost. On the edge of the McGuire property, a half mile from the Empire Mine, is the old tailings dump of the long-since closed Sultana Mine. These tailings assay $12.50 per ton. They were put in their present position when only the free gold in the ore was being saved. All the gold tied up in the cubes of iron sulfide flowed off the end of the mercury table and on into the tailings pond.

The operators of the small mine knew from the assay reports that the gold was being lost. They built a small dam to store the tailings, hoping that some day they would be able to recover the lost values. The vein pinched out and the mine ceased operations before a practical method of reworking the sand was discovered, so the material has stayed in place during all the ensuing years.

Most of the old mine dumps similar to the Sultana's were reworked during the Depressiion years. But in 1918, the Empire Mine had bought all the holdings of the Sultana. The 2,000 tons of the old tailings were now so thoroughly mixed with pine needles and pine tree roots that careful hand work would be needed to make the processing of the sands economical. This the mine did not think was practical, so the sand remained intact until the war years of 1941-45.

During that period, someone found that the sand from the old dump was exactly what was needed to loosen the hard red clay soil in their gardens. The dump lay within 100 yards of Rattlesnake Road. All but a few tons disappeared within a few months. The Empire posted a sign to the effect that this was private property and no one was to trespass, but that only served to point out that something of value was there. No one was ever caught hauling it away.

Shortly after the Sultana Mine closed, Frank Wilfley invented a vibrating table that successfully separated the heavy metallic minerals from the lighter sands. Wilfley's table was well named. He took an ordinary table top, mounted it on an eccentric which caused it to vibrate longitudinally. To the top of the table he nailed a series of wooden cleats, each tapered from one-half inch in thickness on the thicker end, to zero at the opposite end, and each slightly shorter than the one nailed immediately above on the table top. The whole table was then tipped to slope at about 10 degrees horizontally towards the side of the table carrying the longest cleats.

The vibrating table was placed immediately below the copper amalgamation plates under the stamp mill. The water, the crushed sand, the sulfides, and the very fine gold which had not adhered to the mercury, spilled onto the upper end of the table. The vibrations caused the materials to be carried towards the opposite end where the heavier materials were spilled into a gutter and washed into a pool to be recovered. The lighter materials, those with no value, washed over the top of the cleats. Gravity pulled them to the lower side of the table where they too fell off into another collecting trough and were washed out of the mill as almost worthless tailings.

At the Empire Mine the sulfides coming off the end of the Wilfley table were flushed into a small ball mill and ground to a 250 mesh. Then they were sent to the cyanide plant where the gold was extracted by chemical means. Most other mines dried the sulfides and shipped them by truck to the smeltery at Selby, on San Francisco Bay.

The gold was recovered from the sulfides at the smeltery by a process which went as follows: The materials to be smelted were mixed with fluxing materials such as soda ash (calcium carbonate) and put into a large crucible. High heat was applied. This did two things. It fused or melted the mixture and it burned the sulfur contained in the metallic sulfides. The sulfur dioxide formed was allowed to escape through the smeltery stacks.

The metals from the sulfides were now freed and were suspended in the melted flux in elemental form. Powdered lead was then introduced in relatively large quantities. Being heavy it settled to the bottom of the crucible. As it did so, it picked up the very fine particles of any other metals and carried them along.

The flux or slag was then poured off and the molten metal introduced into another furnace where the base metals such as the lead, the zinc, the copper, and all others except the royal metals were oxidized. These oxides were trapped in cooling stacks to be reduced to metals by other processes. The gold, the silver, and other royal metals, if any were present, remained because they cannot be oxidized.

The gold was separated from the silver by simply dissolving the silver in a solution of nitric acid. Silver nitrate (used in photographic films) was formed. The gold did not dissolve. It was filtered out, melted, and cast into bars to be sent to the Mint.

At the Empire Mine, the cost of transporting the sulfides—at least 50 tons a day—and the cost of smelting the ore was saved by putting the sulfides through the cyanide process. That consisted of the grinding of the sulfides into a fine powder. This powder was placed in large redwood tanks, each capable of holding several thousand tons of the heavy material. When the sulfides were in place, they were mixed with a dilute solution of potassium or sodium cyanide.

Compressed air was then forced through the mixture to agitate it and to supply the oxygen needed in carrying out the chemical reaction. Under these conditions, gold will dissolve in the cyanide solution, forming the compound of auric cyanide. (Jewelers use a paste containing potassium cyanide. It removes the tarnish from gold or silver objects by dissoving a very thin coat of the metal from beneath the compounds forming the tarnish. Only the professional may use the cyanide compound. It is highly toxic.)

The combination of carbon and nitrogen (CN) is entirely artificial. It does not occur in nature in commercial amounts. Large amounts of electricity is needed to force the combination of the two elements. Only after great amounts of electrical energy became available at Niagara Falls was it practical to make the compound.

It has one more use that is widely known and

frequently frowned upon. Lumps of potassium cyanide are placed in a dish below the condemned man's chair in the gas chamber of many penitentiaries. When the executioner "does his thing," he tips a bottle containing sulfuric acid, allowing some of the liquid to spill over the cyanide pellets. Hydrogen cyanide is released. One breath of the concentrated gas is sufficient to cause one to lose consciousness. Death results within minutes.

Before the cyanide process was invented, it was known that gold can be put into solution by dissolving it in aqua regia. This is composed of freshly mixed concentrated hydrochloric and nitric acids. The miners knew of this process, but it was much too expensive to use on the thousands of tons of sulfides to extract the gold they contained. Gold goes into solution in the acid mixtures because gold chloride is formed. That is a soluble compound.

Other methods of forming the chloride of gold were known and tried. First, chlorine was liberated from common salt ($NaCl$) by the use of an electric current. Freshly liberated chlorine (nascent) will dissolve gold in an oxidizing solution. In the laboratory and in the pilot plant the process works, but on an industrial scale it was never made to work before the cyanide process became available. Next it was tried to force the combination of gold and chlorine to form the soluble salt by a roasting process. The gold ores were mixed with salt and roasted at a high temperature in a furnace. Soluble gold chloride was formed, but the process proved to be too expensive to be of use to the miners who usually had ores of such low grade that the cost of the process was too great.

The gold attached to the sulfide crystals still defies the attempts to dissolve it all by the use of even the cyanide solutions. The sulfides went into the cyanide tanks at the Empire Mine carrying about $250.00 per ton in gold. Try as they might, the chemists were never able to extract the final $10.00 from each ton of the material treated. The mine management knew this, and hoping that someday, someone would be able to solve the problem, they had the fully treated sulfides piled near the cyanide plant. In all, the Empire Mine piled in excess of 100,000 tons of the material.[1]

Both the Empire and Idaho-Maryland Mines stayed with the ore treatment processes described above until long after most of the mines of the world had shifted to the more efficient methods of ball mill grinding, and oil floatation to separate the values from the "gangue" or waste materials.

Even then, the Empire did retain a few stamps to do the primary crushing. From that point on, the modern method followed by the Idaho-Maryland and its satellite mine, the New Brunswick, was adopted by the Empire.

As the ore came from underground, it was dropped into a jaw crusher which reduced the ore to pieces of no more than 3/4 inch in size. Next it was fed by gravity into a cone crusher. This

---

[1]In 1940, Fred Nobs, the superintendent of the Empire Mine gave the writer $1,000,000! The "Catch 22" was that he had to recover it from the spent sulfides which were plainly visible out of his office window. Needless to say, the sulfides are still there. Only now each ton is worth more than $100.00! When the visitor goes to the recently formed Empire State Park in Grass Valley, he has the thrill of seeing $6,000,000.00 lying out in the open. During the period between the closing of the mine and take-over by the State, the sulfides could have been hauled away by anyone who might have had a use for them.

consisted of two corrugated cones, one fitting inside the other. At the top, the clearance between the two cones was about one inch. The space between the two cones was gradually reduced until, at the bottom of the machine, the space did not exceed one-tenth of an inch. The inner cone was forced to revolve while the outer cone remained stationary. As the ore worked its way down between the cones, it became smaller and smaller. It dropped from the bottom as a coarse powder. In effect, the cone grinder was nothing but an old-fashioned coffee mill on a gigantic scale.

Upon leaving the cone crusher, the ore was fed into a revolving steel cylinder or drum. There it was mixed with water and rolled over and over with steel balls (hence the "ball mill"), or sometimes with very hard and heavy rocks. The crushing and abrasive action of the balls reduced the ore to the consistency of a fine powder, the degree of fineness depending on the volume of water introduced along with the ore at the upper end of the mill. The greater the flow of water, the larger the size of the particles which flushed out the lower end.

This, of course, was a very crude method of grading the ore to the proper size, so a classifier was introduced into the line. This machine picked up the pieces too large for the floatation cell and fed them back into the mill for further grinding.

The properly sized ore, usually 200 mesh or less, was then mixed with the proper chemicals (dressed) and fed into the floatation machines. These consisted of tanks where the "slurry" could be agitated by compressed air and the foam which developed skimmed from the top of the liquid, just as cream was skimmed from a pan of milk by the writer's mother and grandmother. This foam contained all of the metallic mineral compounds of the ore.

The valueless quartz and other rock materials were not coated by the ore dressings. They did not rise to the top, rather they sank to the bottom from where they were drawn off and discarded. Most usually, the floatation cells were combined in batteries of four. The powdered ore was passed through each cell in succession. Thus, each particle of valuable material had four chances of being separated from the bubbling mixture of sand and water. The foamy material was drained into a sump. There it was dried and later sent to be smelted.

The "Idaho" inserted one more step in the process. Sandwiched between the ball mill and the classifier was an old fashioned sluice box. This was used because the ore was often rich in relatively large pieces of free gold. These nuggets were so highly malleable that they were rolled out into large flakes and had a tendency to be carried more rapidly by the water current. This made them easy to recover in the sluice box and eliminated the necessity of grinding them into a powder by repeated passes through the mill.

Needless to say, the sluice box was covered with a very heavy wire screen. The screen was attached to the box by three padlocks. Each lock had a separate key, and each key was carried by a different man—presumably, each was a trusted employee.

Ball mills come in various sizes and shapes. They were not developed in the Nevada County area, as were so many other new mining techniques. That distinction belongs to the copper mining districts where great tonnages of low grade ores had to be crushed and ground to a very fine powder. This the stamp mill could not do.

Likewise, the other great modern innovation in the mining industry, the oil floatation process, was invented and developed to separate small amounts of metal bearing minerals from large amounts of worthless (gangue) material. It too was developed in the copper mines.

The phenomenon of the floatation system, where the heavy metallic minerals float to the surface of the water, rather than to sink to the bottom, as it does in all other separation processes, is supposed to have been discovered by a woman. As far as is known, this is the only major contribution made to the science of mining by a female.

It seems that the good lady's husband was operating a drilling machine in a rich stope in Butte, Montana. He was working in such crowded quarters that his working clothes became covered with the very fine dust from the highgrade copper ore. Periodically he would bring his overalls home to be laundered. Washing machines were not as yet available to a miner's family. An old fashioned tub and washboard was used. The good lady accidentally spilled more of the liquid home-made soap into the tub than she had intended. As a result the brisk motion of the clothes being rubbed up and down against the corrugations of the washboard produced so many suds that they threatened to run over the wooden sides of the tub.

The early morning sun streaming in through the kitchen window reflected from the surface of the bubbles, giving off the brilliant peacock colors common to soap bubbles. Only this time, something more was added. Mixed with the purples and the blues were specks of gold!

In great excitement, the housewife waited for her husband to return from the day shift. She told him what had happened and then added that she had saved as many of the suds as she could. The miner poo-pooed the idea that his wife had struck gold. But he was curious enough to get out a small gold pan and carefully pan the sample of suds, now completely deflated. To his amazement he found that there was indeed a tiny trail of gold colored material in the crease of his pan.

He was too good a miner to be fooled by what his wife had thought to be gold. The answer was obvious. He had been drilling through an almost solid ledge of chalcopyrite (copper sulfide), one of the many metallic compounds of a base metal and sulfur commonly known as "fool's gold."

The ledge the miner was working was of such quality that the rock dust which settled on his clothing contained a large proportion of the yellow copper ore. He had, however, worked in parts of the mine which produced only low grade ore. In fact, he had just recently been moved from a low grade area because there was no known way of concentrating the ore to make the resulting material worth the cost of shipping and smelting. He pondered the matter for some time, then he brought home a piece of the rich ore. This he ground to a fine powder and sprinkled it liberally over a clean set of work clothes. When his wife washed the (salted) clothing in water which contained enough soap to foam copiously, he carefully skimmed the suds off the top of the tub, then dried and panned the results. Again he found the streak of yellow copper ore in the crease of his pan.

Thus was born the floatation cell. The process which, as the miner said, "made gravity work upside down."

Later, when the metallurgists worked with the new method of concentrating low grade ores, it was found that by changing the formula of the "soap," or ore dressing, not only was it possible to separate the metal bearing minerals from the gangue, but they could also separate one base metal from another: iron sulfide from copper sulfide, etc.

During the Great Depression years of the

1930's, one of the favorite methods of the retired miners was to find an abandoned mill in the Grass Valley area which had processed gold ore at one time—such as the Sultana Mine, previously mentioned.

Once such a place was found, the method of making a few dollars was simple. If any timbers remained, they were burned and the ashes panned. Old flooring, in particular, often had significant amounts of very fine gold and gold amalgam embedded in the cracks. All of the soil around the bases of the ore crushers and the stamp mills was then put through a small sluice box and panned. Lastly, if it was possible to find the remnants of the tailing pile, the sulfides of the older mines would be run through a portable floatation cell. The concentrates would then be hauled to the smeltery and, eventually, a check would arrive in the mail.

The recovery of such sulfides, those deposited by a slightly different method, leads to yet another story of the Highgraders.

In 1938, Old Jack Combs passed on to the land where the streets are reported to be paved with gold and precious stones. He left a bereaved widow. Jack had worked all his life in the local gold mines and, according to the custom of the times, he had managed to get out of the rich stopes, where he was blasting down "picture rock," with a few pieces of highgrade in his pockets.

There was a small creek running through his back yard. It made a very convenient place to pan the gold after he had run it through a small "squirrel crusher" and ground it to about 50 mesh in a steel mortar.

Since the creek had a tendency to dwindle to a trickle during the long dry summers, Jack built a small dam across a narrow point in the channel. This not only assured him water to do his panning, it also provided a pool where panning was much easier than in running water and it trapped the sulfides which always occurred along with the free gold he was able to recover by panning. Jack knew that these sulfides were "shot full" of gold, but he had no way of recovering that gold. He hoped someday to be able to get something for the riches he knew to be on the bottom of his pond.

Jack was well known among the Cousin Jack miners, so for a fee he crushed and ground their highgrade and helped them do the panning. This went on for years. At last the rheumatism Jack had acquired by working in the cold and the damp of the mines made life too miserable. He retired. He had a few investments in gilt edged stocks and he continued to help his friends recover the gold from their highgrade. He and his wife lived rather well.

Jack passed on before he could figure out a way to get his money out of the sulfides at the bottom of the pond. His widow realized that she could never do anything with the gold she knew to be part of Jim's estate. She called Mike McGuire and asked him to come right down. He came, hat in hand, and after the proper condolences, asked what he could do for her. She took him to the creek and pointed out the sulfides on the bottom of the pond.

Mike knew very well what was there. In fact he had sent many a small highgrader to Jack when he had been approached to help get the gold out of their rock and on its way to the Mint.

Mrs. Combs figured that since Mike was an Irishman, and the Irish were always at odds with the Cousin Jacks (unless there was highgrade involved), it was safer to let Mike buy the sulfides than to call in some of the "Cousins." She guessed that a hundred dollars would be about right to give her a fair return and to allow Mike a small profit on his investment. Mike figured "as how" he could not invest a hundred dollars with a better chance of making a good profit, so he took the offer.

Mike filled twenty carbide cans with the scrapings from the bottom of the pond and stored them in his garage. The cans filled with the sulfides stayed there until after the war. In 1948, Mike figured that it was about time he found out what he had bought, so he took an auger and bored out a sample from each of the cans. He then took a composite sample of the borings to the American Smelting and Refining Company at Selby.

At the office of the smelting company, he was met with disinterest as he had explained that he had only about one ton of the material, but when he told the superintendent that the concentrates were the result of cleaning up around an old mill in Grass Valley (he failed to tell them what kind of a mill it had been) the smelting company suddenly got very interested indeed. The assayer informed him that he would have a report on the value of his concentrates within the week.

In due time the assay report came through. The concentrates were worth one dollar eighteen and one-half cents per pound in gold and fourteen cents in silver. Mike hauled the twenty cans to the smeltery in his pickup truck. The check for the concentrates, after smelting costs and assaying fees had been deducted, came to exactly one thousand eight hundred fifty-six dollars and twenty-four cents! Mike's one-hundred dollar investment had turned into a small gold mine!

CHAPTER XXII

PELTON'S WHEEL, OLIVER'S FILTER, AND THE CORNISH PUMP

Among the more important inventions made in the gold fields of California was the little-known Pelton Water Wheel. Little known beyond the generation of electricity, but so much used in that field, that since the turn of the century it has produced more wealth in the mountains of California than all of the gold ever dug from the river beds and from the deep quartz mines. It forms the basis for the production of the hydroelectric power generated and distributed by the Pacific Gas and Electric Company. In fact, it was Pelton's invention which led to the formation of this, the largest of all privately financed public utility companies.

Until Pelton's wheel was put to work, only two types of waterwheels were in common use, the overshot, and the undershot. The overshot wheel was used where the volume of water was relatively small and the drop relatively high, as at the base of a small waterfall. This wheel was constructed so that the falling water entered a series of troughs or buckets, fixed to the rim of the wheel in such a way as to cause the weight of the trapped water to pull the containers forward and down. When the troughs or buckets had reached the bottom of the 180 degree arc, the water spilled out, and the empty containers continued up the reverse 180 degrees back to the top. Here each trough was refilled by the supply of running water. The process was repeated, causing the wheel to rotate continuously. Power was taken from the revolving axle by means of a pulley and belt or by an arrangement of cogs and gears. This, of course, was a most inefficient method of converting the force of falling water into useful work, but the only improvement that could ever be made was to reduce the friction on the moving parts.

The undershot wheel was used where the volume of water was greater, but the fall was relatively small. This was the type of waterwheel being installed by Marshall at Coloma when the first bits of gold were picked out of the mill race (the channel which carried the water from the American River to and away from the mill wheel).

The undershot wheel is made by fixing a series of paddles to the ends of the spokes of a wheel. These paddles are submerged in a stream of rushing water. The momentum of the moving water pushes the paddles ahead, causing the wheel to spin. Fulton used the same principle on his paddle wheel driven steamboat, except that he forced the wheel to turn by means of a steam engine. The inertia of the water behind the paddles pushed the boat ahead.

As with the overshot wheel, the efficiency of the undershot wheel could be improved by reducing friction. In addition the rushing water can be confined in a channel, the sides of which fit more tightly against the outer edges of the paddles. This causes it to give up more of its kenetic energy as it forces its way down the race. Other than these improvements, nothing has ever been done to improve the wheels since the dawn of written history.

Lester Pelton was a Yankee "jack of all trades" from Ohio. He was working for the great hydraulic mines on the Middle Yuba River near Camptonville. He was fascinated by the tremendous power of the water as it rushed from the throats of the monitors. It reduced the seemingly solid rock to loose sand and gravel. He reasoned that if he could convert this energy into useful work, he could accomplish things which all other types of waterwheels could not do.

In addition, the hydraulic mines were about to be closed by the irate ranchers in the valley below. They had just succeeded in having laws passed in Sacramento making the washing of hydraulic slickings onto their land illegal. The mines were about to close and the workers would have to look for other employment.

He also had an immediate and practical need for a machine which would convert the energy of the water spurting from a tiny pipe which ran from a spring on the hillside above his landlady's house. The good lady recently purchased one of those "new-fangled" treadle driven sewing machines; but, her rheumatic knees made it impossible for her to operate the machine in the normal fashion. She begged Pelton to work the treadle for her. He could not refuse, but being a man, he also could not be caught by the other miners, doing a woman's work. He was forced to figure some way out.

He borrowed some tools from a friend who ran a blacksmith shop. With these, he built a tiny waterwheel. He constructed it exactly like the undershot wheels which worked in the river below. He had no running water to submerge the blades in, so he mounted his miniature wheel so that the water from the spring struck the lower blades of the wheel as it squirted from the pipe. The wheel was attached to a post by a shaft and a bearing. He attached a pulley to the opposite end of the shaft. A small belt was run from this pulley to the sewing machine.

The apparatus worked just as Pelton had visualized. He no longer had to help his landlady. He was, however, disappointed with the amount of power produced. He increased the efficiency of his wheel by visiting the garbage dump at the rear of the Camptonville Hotel. There he found a number of empty oyster tins. These he fastened to his wheel to replace the paddles. He reasoned that if the water was contained in the cup-like tins, more of its energy would be trapped. He was right to a degree, but the efficiency of his wheel was increased only slightly. The trouble was very evident. The rushing water struck the paddles or the cups and bounced straight back into the path of the on-coming stream. Most of its force was lost by this interference.

Supposedly, Pelton solved his problem after observing his landlady trying to drive a stray cow out of her garden patch. [It is remarkable how many times a cow has received the blame for, or has been credited with some tragedy, or with some discovery or invention; witness the Chicago Fire.] The stubborn bovine refused to leave the succulent corn and beans. In desperation, the lady picked up the garden hose and played the full force of the water into the cow's face.

Pelton laughed at the antics of the cow as she tried to escape the unexpected washing. Then he sobered and practically ran back to his small shop where he still had the blacksmith's tools. He had noticed that when the stream of water hit the nose of the cow, the cartilage between the nostrils divided and threw it to either side. None of it bounced back into the on-coming stream.

Pelton modified the cups on the rim of his waterwheel. He made the outer edges slope away from the center of the wheel rim, and he made a sharp ridge at the center of each cup. He then fastened the cups so that the stream of water struck the ridge of each cup "dead center." When he turned the water against his new wheel, the unexpected happened. First, the sewing machine began running so fast that its vibrations threw the belt off the pulley. Next, the wheel began running so fast that it began to "sing" and then to "whistle." Finally, with reports like rifle shots, the new cups started popping their rivets and were sent flying against the roof!

After it was too late, Pelton got the water turned off and sat down to contemplate the wreckage. In spite of the disaster, he was highly elated. He knew that he had solved the problem. He had only to perfect a means of controlling his wheel, and he would have let a genie out of a box; only this time, it was out of a pipe.

Pelton moved to Nevada City, where James Allen, owner of the Allen Foundry, tested the new wheel and entered into a working agreement to manufacture the new device.

The deep quartz mines in western Nevada County in late 1878, had already succeeded in denuding all the surrounding hills of their standing timber to feed the steam boilers which were supplying power for the needed compressed air. Water under high pressure from an abandoned hydraulic mine near the Idaho Mine was available. Allen and Pelton attached a new and larger wheel to the mine's air compressor. To the amazement of the mining and the scientific world, 85% of the kenetic energy of the falling water was converted into compressed air useful in the mine.

Immediately, every other mine with a source of falling water ordered one of the new wheels. The Department of Mechanical Engineering at the University of California tested the invention under controlled conditions. It found that Pelton's calculations were correct. [The Pelton wheel is still the second most efficient machine ever to be built. Only the modern water turbine is more efficient. It is capable of producing an efficiency of about 92%. But it is only of use where the amount of water is very great and the fall relatively small.] Pelton wheels are still being manufactured and installed the world over for the production of hydroelectricity. The P.G.& E. Company installed a very large one as late as 1977 at its power house at Caribou on the Feather River.

The largest diameter wheel ever built was constructed under license, by the North Star Mining Company at the foot of Mill Street in Grass Valley. The wheel is still in place. It is now a part of the North Star Mining Museum maintained by the Nevada County Historical Society. It is worthy of note that this wheel was designed and fabricated by the Superintendent of the North Star Mine, Arthur B. Foote. Mr. Foote and his wife were the chief characters in "The Angle of Repose," Wallace Stegner's Pulitzer Prize novel, Doubleday, 1971. In the novel, the Footes were the "Wards."

In 1890, Pelton, with the help of an electrical engineer named DeSable, and with the finances of Colgate, a son of the founder of the Colgate Soap Company of Cincinnati, attached a Pelton Wheel to an alternating current generator by means of a direct drive and modern electrical generation and transmission was born. The deep quartz mines were the first industries to adopt this convenient form of power. Eventually, the Great Western Power Company delivered power to the Bay Area cities, and to all of northern California. In 1924, that company was combined with the Pacific Gas Company of San Francisco, and so was born the P.G.& E. Company (Pacific Gas & Electric Co.).

With the development of the cyanide process for the dissolving of gold from the finely ground quartz and sulfides, came the need for large scale continuous filtering. A mill man named Henry Oliver saw this need and did something about it. He solved the problem of separating the gold bearing solution from the fine sands in order that the gold could be precipitated and the useless materials discarded.

From a normal mixture of sand and water, the water could be drained off by putting the mixture in a container with a canvas bottom. The sand from the cyanide tanks, however, was so fine that it formed an impervious layer that allowed no water to pass.

Large filter presses were tried, but the amounts of material to be handled were so great that these were impractical. The best the mines could do was to allow the mixture of sand and solution to stand for several days, siphon off the liquid, then shovel out the sand in preparation for the next batch. The sand was still saturated by the gold bearing liquid. It must be rinsed several times with fresh water before it could be discarded. This was, of course, a long, slow, costly process.

Oliver's invention consisted of a hollow drum made from very heavy steel screen. This he covered with a heavy canvas. The ends of the drum were solid, except that the drum revolved on a hollow axle. This in turn was connected to a vacuum pump. The drum was submerged to about one-quarter of its diameter in a vat containing the materials to be filtered. As the drum turned, the liquid was drawn through the canvas. The solid material adhered to the surface and was held there until it had passed the top of the arc of the revolving drum. At that point a long knife, held only a few thousandths of an inch above the surface of the canvas, scrapped the now nearly dried material off and let it fall into a trough to be carried away.

The gold-bearing solution was pumped out through the hollow axle of the revolving drum along with the air drawn out by the vacuum pump. The gold, in the form of $AuCN_2$ was recovered by replacing it with zinc dust. Thus gold cyanide plus zinc produced zinc cyanide plus gold.

Oliver's filter was an immediate success. Its use was not limited to the field of mining. It is used the world over in the manufacturing of unknown numbers of chemicals, and in the processing of many of the food products sold in every supermarket.

One of the limiting factors in the development of the early quartz mines was the water which made every mine eventually turn into a well. This water must be removed else all work must stop. The early miners had no method of doing this. Ordinary suction pumps will lift water from only a depth of 32 feet, and bailing out hundreds of gallons per hour with rope and bucket was, of course, impractical. The "Cornish Pump" helped solve the problem.

The prototype of the Cornish Pump was brought to California aboard a clipper ship via Cape Horn. The first of the Cornish miners (Cousin Jacks) who arrived in the Northern Mines saw the need for an efficient system of keeping the mines dry. They immediately had the mine owners send to England for the metal parts. When these arrived, they used local timbers to complete the building of the relatively simple machines.

Basically, the pump consisted of a very long shaft operating a plunger in a cylinder located at the bottom of the mine. The plunger forced the water to the surface through a pipe, or in the case of a mine with multiple underground shafts, to reservoirs on higher levels where it could be picked up by another pump cylinder and brought to the surface.

Steel rods for the plunger shafts were not available. That presented no problem. Since the time of the Romans, the Cornish tin mines had been kept water free by the use of wooden beams where today steel rods would be used. To supply the tremendous strength needed by the plunger shaft which often reached hundreds of feet in length, the wooden beams were constructed of six, two by twelve inch fir planks. Staggered lengths were used. These were bolted together to form a continuous timber of laminated wood. Surprisingly, the twelve by twelve inch timber was quite flexible. This made it possible for the long pump shaft to follow the dips and angles of the mine shaft as it followed the ledge deep into the interior of the earth.

The power plant to drive the machine was located on the surface by the collar of the mine shaft. A powerful steam engine was used to drive the gears which converted the rotary motion of the flywheel into the perpendicular motion of the pump plunger. In operation, the gear and plunger action looked like the "walking beam" used to activate the side wheel paddles of the old Southern Pacific ferry boats on San Franciso Bay. Walking beams are familiar to anyone who has ever driven by or through a pumping oil field. The rocking arms of the oil pumps are actually small walking beams.

The Cornish Pumps kept the water out of the deep California mines until the high pressure centrifugal and rotary pumps were developed. Later they also became the basis for the deep well pumps used in the citrus groves in southern California, where cylinders had to be lowered to a depth of two thousand feet or more to reach the needed irrigation water. By that time, however, steel of sufficient tensile strength to withstand the enormous pull had been developed so that small steel rods could be used.

As the mines were sunk deeper, the amount of water to be raised to the surface became greater. By 1930, the Idaho-Maryland Mine was pumping 20,000 gallons daily from an average depth of 2,500 feet. Of course, no one pump "forked" it all the way up in a single step. Reservoirs were placed at 500 foot intervals, but even at those depths the water pressure built up to more than 200 pounds per square inch, so highly efficient pumps, as well as very powerful ones, had to be used.

Strangely, the Cornish pump was used in a few of the Nevada County mines until near the middle of the 1900's. The last one of record was at the Sneath-Clay Mine on the edge of the city of Nevada City. The mine was abandoned at the end of World War II when labor costs began to escalate. The mine buildings collapsed under the weight of the winter snows. The machinery stood exposed to the elements. In 1978, the Nevada County Historical Society salvaged this, the last known example of the famous pump. They dismantled it and hauled it to their museum located at the foot of Mill Street in Grass Valley. There it was reassembled. The operational machine is on display along with the North Star-Pelton Water Wheel. The old and rotted timbers were replaced by new.

It is of interest to note that Herbert Hoover had the responsibility for the operation of this pump when he took his first position as a mining engineer following his graduation from Stanford.

The mines went to great lengths to trap as much of the underground water as near the surface as possible. The cost of keeping the mines dry increased in geometrical proportion as the mine deepened. That fact leads to yet another human interest story which leaked from the mines in 1938.

Bill Jones (fictitious name) got his degree in mining engineering in 1936. He was employed by the Empire Mine in Grass Valley as their underground surveyor. He was paid $250.00 per month, a relatively high wage for a boy just out of college at the time. But he was not satisfied. That amount of money did not let him live in the manner he had been accustomed to at the Eastern university he attended.

To improve his lot, he approached the management of the Scotia Mine, which only months before had purchased the Golden Center Mine along with all its old underground workings. Bill told the management of the combined mines that if they would employ him he would guarantee that they would be able to shut down their pumps and reduce their power bill by more than half. In return, they were to double his salary. He also told them that if he could not deliver, they need pay him nothing! Naturally, he was employed on the spot.

Bill made a careful survey of the old Golden Center workings. Then he took a diamond drill crew underground and started them drilling at the lowest point of the combined mines. Just as his calculations had shown, the diamond drill, after only one hundred feet, broke into an old stope of the Empire Mine. All the water from the Scotia Mine disappeared down a "rat hole"!

For some months the Scotia Mine paid nothing for pumping. But the "joke" on the Empire was too good to keep. Some miner got drunk at the Bret Harte Bar on a Saturday night and talked too much. Every miner in town had a good belly laugh at the thought of the little struggling mine outwitting the greatest mine in California.

The Empire, of course, did not think it funny at all. Search as they would, they had not been able to find the source of the sudden rush of new water into the 3,000 foot level of their mine. Their pumping costs from the deeper parts of their mine had suddenly more than doubled. They immediately brought suit against the Scotia. Naturally they won. They collected all of their increased power cost, plus punitive damages as well.

Bill? Oh, he was fired. But that did not worry him. He was immediately hired by the Lava Cap Mine. Anyone that smart would just have to be an asset to any mine!

That water pressure underground became great and could be very dangerous if not handled correctly, was demonstrated in 1950, when the State Division of Industrial Safety did not allow the New Brunswick Mine to crosscut, on the 5,000 foot level, to pick up the ancient workings of the Union Hill Mine. The mine had been losing money in trying to mine gold with the price fixed at $35.00 per ounce.

Records had shown that the Union Hill Mine had closed in the mid 1920's because the gold in the ledge they were following had suddenly been replaced by wolframite, the ore of tungsten. When found, tungsten was relatively cheap. Now, following the War, it had become more valuable, and the government was paying a premium to have it produced in this country. The New Brunswick believed that it would pay to mine the ore if they could get at it from their own 5,000 foot level.

The State Safety Division ordered the crosscut stopped because at that depth the water pressure in the old mine would be more than 2,000 pounds per square inch. A drill hole into water under that amount of pressure would produce a stream of water so hard that it would pass right through a man's body. Or if there was a sudden breakthrough, the rush of water from the old workings could flood the lower reaches of the New Brunswick, with the resulting loss of many lives.

When the Empire Mine stopped operations in 1957, they dammed off the drifts leading to the underground workings of the Pennsylvania and the North Star Mines. They did this in order that, should any one of the mines open in the future, the other mines would not have to be de-watered along with the area to be worked. The engineers figured that to withstand the pressures at the lower depths, reinforced concrete plugs up to thirty feet in thickness would have to be poured in the tunnels if they were to hold with safety when the water was pumped from the other side.

Almost a full year was required for a crew of twenty-five men to "pull" all of the mining machinery from underground, once mining had stopped, and to pour the concrete needed to plug all the drifts.

As mentioned previously, the used machinery, some more than forty years old, sold at second-hand prices for more than one and one-half million dollars!

Underground Mule Train (Empire Mine)

Rocker & Panning, Placer Mining. Depression Miners, 1930-1940. Reworking Dry Creek in the Spring.

— Photo Courtesy of California State Library

CORNISH PUMP. RESTORED AND OPERATIONAL

MINE HOIST. 2,500 Ft. 3/4 " Cable

CIRCA 1850-60. No Hard Hats. No Carbide head-lamps.
MAN CARS. EMPIRE MINE. UP CAR RIGHT

MOTOR AND GEAR ASSEMBLY. CORNISH PUMP

ROCKER ARM, PISTON AND SUCTION PUMP
TWO INCH PIPE GOING TO WATER.

CHAPTER XXIII

HIGHGRADE AND THE HIGHGRADERS

During the first two decades of the Gold Rush, gold dust was an accepted medium of exchange. It passed from hand to hand either by the "pinch," when amounts of a dollar or less were involved; or by the ounce, weighed out on the grocer's scales or on the balances in the Wells Fargo office. A full pinch equaled a dollar; and a Troy ounce was valued at from sixteen to eighteen dollars, depending on the district where it had been mined and the avarice of the gold buyer. The only method of getting the full value of the newly mined gold was to ship it to the Mint in San Francisco. There it was assayed for both its gold and its silver content. That being known, the government then was able to pay the owner its correct value of $20.67 per fine ounce plus the value of the silver.

After 1870, paper currency and gold coins came into general circulation. Only the stores in the isolated mountain towns continued to accept dust in payment for goods and services. At that point the professional gold buyer became a necessity for the prospectors coming out of the hills, but having no time to wait for a return from the Mint.

At first there was no question as to the ownership of the gold offered to the buyer. Possession proved ownership. In fact, it would not have been healthy to question the man straight out of the hills and the desert as to where he got that poke full of dust. It was plain to see that his one thought was to get some spendable cash and into the nearest saloon to do some serious drinking. Men were shot for less!

When the amount of gold brought in by prospectors began to dwindle, the gold buyers, as a group of semi-professionals, looked for means of continuing their easy source of income. The answer to their problem was immediately at hand. Most, but not all, of the buyers discreetly let it be known that if the men working in the mines operated by the large companies would let them handle the highgrade they knew was being carried away, they, the buyers, would pay top prices! Thus the legitimate gold buyer became a "fence," except that he also was known as a "highgrader."

In addition to the professional gold buyers, others took advantage of the miner's need for turning illegal gold into cash. Right up to the time of the closing of the mines in 1957, there were always a number of small mines surviving by becoming a means of getting rid of highgrade. A prospector with a small pay shoot which he was sure would turn into a bonanza if he could hold out just one month more was an easy mark for the highgrader with fifty or more ounces of gold to get rid of. For five dollars an ounce, the real miner would mix the illegal gold with his next shipment to the Mint. In due time, if the Treasury men were not suspicious, he got his check, cashed it, and paid the highgrader. The two hundred fifty dollars he got for his part of the transaction was often the difference between keeping his mine open and operating and closing down. Of course, the easy money acted exactly as a drug. He became addicted and like the addict, he continued the operation until his own mine started to pay "big," or he was caught.

This method was used by some relatively large mines in Utah and in Nevada. The larger the mine, the larger their own percentage of the gold and silver shipped, the harder it was to prove duplicity. Eventually, however, most of them were caught. That happened because someone talked. Once the Mint had been alerted, it was usually easy to prove that some of the gold did not come from the mine which had shipped it. Just where it did come from might be harder to prove, but that was not necessary. The fact that the shipper had sworn to a false affidavit was enough to get him into big trouble.

Most people think that "gold is gold," and that one piece of the yellow metal looks just like any other. They could not be more mistaken. The Mint man receiving the gold in San Francisco could tell at a glance which mine a Grass Valley shipment came from. The Grass Valley miners could do even better. They could tell from what level of each mine! If a bar or two of Empire Mine gold showed up in a Toulumne County shipment, the Mint spotted it immediately. The Toulumne mine might try to disguise the northern gold by melting it with their own, but in doing so theirs lost its own peculiar characteristics. The Mint would known that something was wrong.

The highgrade buyers preferred to deal with a California mine in finding a way to get the illegal gold to the Mint. That was especially true before gold was nationalized in 1933. Before that date, stolen gold was a state matter only. It was almost as hard to get a conviction of a highgrade buyer in the local courts as it was to convict the highgrading miner. Before the nationalizaton of gold, once a state line was crossed, stolen gold became a case for the federal courts. There, well away from the mining country, the jury would be less sympathetic. Good hard evidence usually brought a conviction.

After 1933, every case became a federal case. Convictions were much easier to obtain. Some gold buyers actually went to jail! In addition to the federal intervention, the state also stepped up its efforts to protect the mine owners. However, the inspectors employed by the State Bureau of Mines all too often turned out to be "reformed" highgraders. Who better would know all the tricks and who better would know what to look for?

Once on the job, it was very easy for the investigator to "shake down" the highgrader and his fences. For a fee, they were in a position to declare that the gold coming in from a questionable source was legitimate. Thereafter, all shipments from that source need have no fear of confiscation so long as the inspector got his cut. One or two of the inspectors on the make during the Depression years were caught and convicted. They spent some time in a federal prison. Mostly, however, the dishonest state men retired early and lived "the good life." (The March 1936, Mining and Industrial Journal, carried a story of the arrest and conviction of a dishonest inspector. The case was so unusual that it warranted a full feature article in the magazine.)

As told previously, W.C. Way was "grub-staked" in 1935 to open and operate a small quartz mine near Fort Jones in Siskiyou County. The gold in the small vein he was following happened to be of unusual fineness. It assayed 950 fine. That is, 950 parts in every 1,000 was pure gold, the other 50 parts being mostly silver. Most of the gold mined in the northern tier of counties in California assayed between 750 and 850 fine.

Way was not surprised when a state inspector showed up about ten days after his first shipment of sponge gold reached the Mint. The state man went into the mine and took his own sample from the vein which Way pointed out as being the source of the gold he had shipped. He made sure that he took his samples from deep enough below the working face of the ledge so that there was no way it could have been salted. The state then had its own assayer run the samples to make sure that the gold from that mine was indeed of such high quality.

If there had been a change in the purity of the gold in that small distance in the ledge, Mr. Way would have been in serious trouble. This would have been doubly true because it was known that one of his grubstakers had, at one time lived in Grass Valley where highgrading was a major problem and where the gold assayed from 900 to 950 fine. He, Way, was quite worried for about ten days. Then the check for the shipment came through. Thereafter, no questions were asked, and the following Mint receipts were paid without delay.

It is of interest to note that Mr. Way might not have been in as much trouble as he had thought. In 1939, the particular inspector who had visited his mine retired. He built a spacious home on a hilltop overlooking a portion of the Mother Lode mining country well away from the Northern Mines where it had been his duty to see that the miners and the gold buyers were kept honest! The luxury of the inspector's home could not have been justified by the salary the state paid!

The readers who lived through the latter part of the Depression will undoubtedly remember the quarter- and half-page ads in the Metropolitan Sunday papers, begging people to bring in their old watches, rings, dental bridges, and/or jewelry of any sort, as well as any objects of art which might have gold in them in any form. They assured the seller of the highest prices for the gold in the articles they might like to sell. House to house buyers representing these companies made sure that no one who might possibly have some old gold would be overlooked.

Some of these firms were undoubtedly legitimate, but the most of them used their purchases of old jewelry as a cover-up for sending highgraded gold to the Mint. After all it was easy to show on their books the purchase of 25 twenty-karat gold watch cases, when actually the old cases were only "gold filled." A slight change in their records allowed them to send about 15 ounces of gold to the Mint, all of it apparently legal. The same could be done with old gold watch chains, lockets, etc. It is remarkable how many old fashioned nugget watch chains were dug out of the trunks in the attics all across the country. Chains which had been brought back from Alaska by Uncle George in 1900. According to the official records that is.

Most of the detective work of the Treasury Men and the State Inspectors was not of the cloak and dagger type people might have imagined. In a few cases, undercover men were sent into the mines to spy on, and to try to build a case against the miners and the mill men when they observed them highgrading. The majority of the work, however, consisted of routine auditing of the books of the mines, especially those of the smaller mines. They also gave special attention to the records of the large-scale buyers of old jewelry. By comparing the assay reports of the ore being mined and the records of the number of tons mined with the receipts from the Mint, it was usually easy to tell if everything was "on the up and up."

A certain amount of faking of the assay reports and the tonnage milled took place and was undetected, but the inspectors were not dumb. If they were honest, they usually caught the flagrant cases of altered records. If they were not honest, they looked even harder for the false reports. Then they demanded their cut, "or else"!

In 1933, when it became apparent that the price paid for gold by the U.S. Treasury was about to be increased, one relatively large mine in Amador County tried to pull "a fast one," not by mixing highgraded gold with their own shipments, but by holding back some $250,000 in gold bullion instead of shipping it to the Mint as the law stipulated. When the price did advance from $20.67 to $35.00 per ounce, the mine shipped the "old gold" as having been mined after the advance in price. They stood to have a profit of about $170,000 had their scheme succeeded. When they were reported by a miner who had just been fired for pocketing some of the mine's gold, the $250,000 was confiscated and they were fined an additional $250,000. The stock holders missed several of their regular dividends!

With the adoption of the mechanical feeder, the opportunity to pick pieces of highgrade from the stamp mill was lost, but the mill workers were not stopped. They merely shifted from the raw gold in the rock, to the amalgam in the mill. There were two sources of this. One was the amalgam in the stamp chamber, the other the copper-topped table below the stamps.

The mine managers and owners were not stupid. They knew that a certain amount of "nipping" was going on, and while they disapproved, they closed their eyes and allowed it to go on, especially from the mill. They looked with a certain amount of tolerance on the miners' belief that their pay was so low that they were only bringing their income up to a living wage. There undoubtedly was a great deal of justification in this. The mines were paying skilled, experienced miners and mill men $5.00 per day, when carpenters, for example, were getting $8.00 in the same community.

As long as the miners and the mill men kept their highgrading down to a nip now and then, the owners did not press the issue. They knew that if they fired a man for helping himself to a little amalgam from the mill, the man who replaced him might not be satisfied with only a nip. The new man might be greedy. Then the losses could well be considerable. They also knew that it usually did no good to have the highgrader arrested and charged with theft.

From time to time, a miner or a mill man would get caught carrying the evidence off the mine's property. Usually the gold was confiscated and the episode forgotten; but if this particular culprit had been spending too much on fast cars, too much at the local bars, and living beyond the means of an ordinary mine worker, they might try to make an example of him. Almost always they failed. The defendant always demanded a jury trial and it was impossible to get a jury in any of the mining towns without at least one of the jurors being a highgrader himself. If one of the jurors was not actually highgrading, then at least half of the jurors had relatives who at one time or another had been guilty of taking a little gold from one of the mines. A smart defense attorney would always have his client plead "not guilty," and use the plea that he had been "framed." "The shift boss wanted to get rid of him, he planted the evidence!" He almost always got a "not guilty" verdict, or at the very least, a hung jury.

In 1906, Sam Chapman was caught coming off shift with his lunch bucket filled with twenty pieces of exceptionally rich highgrade rock. The amount involved was so large that the Mine decided that an example must be made of Sam. He was arrested and brought to trial.

When the District Attorney submitted the rock as evidence, some of the jurors confessed to poor eyesight. They asked that the highgrade be passed around. They needed to see that it indeed contained gold. The bailiff gave the bucket to the jury foreman. He gravely examined each piece and passed the bucket on to the next man. This went on until the twelfth man had "hefted" and looked closely at the contents of the lunch bucket. He then carefully closed the bucket, nodded his head as if to say that the evidence was indeed convincing, and handed it back to the court official.

The bailiff weighed the bucket in his hand for an instant, then he called for a quick conference with the judge. Together they opened the bucket and found that twelve pieces of the gold laden rock had disappeared!

The judge looked at the jury for a moment, then said, "Gentlemen, I'm going to pass this bucket back around. You better believe me; I'm going to count the pieces when I get the bucket back, and all twelve of those missing pieces had better be there, or God damn it, the whole bunch of you are going to jail for thirty days for contempt of court!"

Needless to say, the jury's verdict was that Chapman had been framed.

Until the automobile became common, the highgrading problem was not too critical. Certainly, every now and then someone did make a big haul. Sometimes he even got away with his larceny. Witness the time that a large mine in Amador County lost a small fortune during the time that Kurt Barr was the superintendent.

Barr had been educated in England where he picked up many of the traditional English customs such as dressing for dinner each evening and taking a brisk ride each morning. One bright spring morning in 1886, Kurt was taking his usual morning ride through the park-like foothills near the mine. Suddenly he was passed on the trail by a gorgeous woman on a horse which was obviously out of control. Kurt, being the gentleman he was, put spurs to his mount and, after a breakneck dash, caught up with the damsel in distress. He reached out and grabbed the unruly animal by the bridle and with a firm hand succeeded in bringing it to a stop.

Linda—Miss Linda Lee—was most grateful.

She half fell from the saddle and collapsed on the ground for a minute, with just the right number of tears and semi-hysterics. Then she controlled her trembling and her tears and threw her arms around Kurt's neck and expressed her gratitude for the brave rescue.

Kurt had never been that close to such charm and beauty before. True, Mrs. Barr was a charming, educated woman, but a woman in her mid-forties, living in a small town in 1886, did not effect even a little bit of glamour. Her time was taken up by the church and by the literary society. She would never think of using even just a tinge of lipstick, nor the touch of rouge worn by this radiant creature, and the faint aroma of the expensive French perfume was the first Kurt had experienced since his school holidays in Paris.

To say that Kurt was smitten would be a gross understatement. The next morning Kurt kept his horse to a walk, but his heart was soon pounding faster than the hoofs of the lady's mount as it cantered up to his side.

Yes, he rode along this path almost every morning. No, he would not be hampered by having a mere woman ride along with him. No, he was not familiar with the New York stage, so he was not in disgrace for failing to recognize the star of "The Great Pretender." Yes, she was going to be here for some time. She was completely exhausted from the long run on Broadway. She just had to hide out in this marvelous place until she was completely rested. Her father had been here in the fifties. He had taken out some gold, but he had tired of the rough life of the miners and had returned to the East where she was born and raised.

The morning rides soon became the great event of each day. She had never expected to meet such a charming gallant gentleman in this isolated place. She was completely captivated. Of course he was married and had a nice family, but after all, she was an actress, and actresses were, by reputation at least, supposed to be quite naughty. Also, it was still the day of the double standard, so Kurt's conscience did not bother him too much when it became almost a daily routine for them to dismount and spend a few delightful minutes "resting" from the hard ride on the carpeting of pine needles in a secluded glade.

After about six weeks of these delightful trysts, Linda appeared depressed on their morning excursion into the woods. Kurt was all concern. He was sure that he could help if she would only confide in him. After exactly the right amount of urging, she confessed that her trouble stemmed from her brother. He had always been a problem. He had just lost his position again—his fifth in only two years. In addition his health was not good. He had to be careful as to what kind of a job he could take.

Kurt laughed at her concern. He told her to wire her brother to come to California immediately. He had a job for him in the mine office. Of course, Ben, the brother, when he arrived, proved to be hopelessly incapable of handling the paperwork of the office. It took a good man a month to straighten out the mistakes he had made in a week. He could not work underground because of a spot on his lung, nor could he work in the mill because of the dampness there. He was not a machinist nor a blacksmith, so he could not work in the mine's shops. There was only one place for him. He went to work in the assay office.

As in most mines, the assay office did much more than test the ore samples taken from underground, as well as analyze for the gold content of the crushed ore going into the mill and of the tailings as they left the mill. Among other things, the workers in the assay office retorted the mercury from the amalgam and melted the sponge gold to be cast into bars of gold bullion to be sent to the Mint.

Ben worked for six months in the assay office before the company detective traced the shortage between the weight of the amalgam coming in from the mill clean-up crew, and the weight of the gold expected to be reclaimed from it. No one other than Ben could have been responsible for the more than $50,000 which was missing.

This was too much for even the superintendent involved with the man's sister to overlook. Had it been a smaller amount, he would merely have fired the man and would have forgotten the whole thing. In fact, he offered this solution to Ben if he would just return the gold. Ben only laughed and reaffirmed his innocence. Kurt had him arrested. He believed that because the amount was so large, and because the thief was an outsider, he could get a conviction.

The following morning Linda failed to show up for the usual morning ride. Kurt knew that she would be very upset. She had seemed to be very fond of her wayward brother and had often told him of her worries about his troubles. She was his only living relative and, as his older sister, she felt a great deal of responsibility for him.

Although they had agreed not to be seen together in public, Kurt was not too surprised when she came to see him in his office. He asked her to sit down and told her how sorry he was that this had happened, but that his hands were tied. Ben was really guilty. He had been caught leaving the assay office with the fraction of the gold bars left at the end of each melt. This was supposed to have been kept in the company safe until the next melt and shipment. He again made the offer that Ben give back the gold, and he would see that he was let off with a suspended sentence.

Linda was more than upset. She was very angry indeed. "Mr. Barr," she snapped, "I demand that you withdraw your charges against Ben immediately." Kurt was hurt by her attitude. He assured her that the matter was out of his hands and that Ben would have to take the consequences for his stealing if he did not return the gold.

Again Linda snapped. "Mr. Barr, you had better drop your charges against Ben, because, you see, Ben is not my brother. He is my husband!"

What chance did Kurt have with a wronged husband on the stand pleading a frame-up to get rid of him? The charges were dropped. Linda and Ben enjoyed their leisurely trip back to New York by way of Hong Kong, Suez, Paris, and London, before looking for still another opportunity to impersonate the famous actress and her unfortunate brother.

Highgrading on a large scale was usually not too common before the advent of the automobile and, especially, not before World War I. It took a great war to get many of the miner's sons out into the world and expose them to a whole new outlook on life. Many of them were no longer satisfied to work out their lives in a "company town," or if they could not escape, they, at least, were not going to be satisfied with the near-starvation wages paid by the mines.

The automobile did even more; it got the older Cousin Jacks, the Italians, and the Irish out of their provincial setting, down to Sacramento, and on to San Francisco, there to see how the rest of the world lived. Their change in attitude was not a change in their morals. Almost every miner would take a little highgrade if the opportunity presented itself. The difference was now that their horizons had been widened, it took money to satisfy their wants. It took a very strong character indeed to pass up a handful of gold amalgam when that new Ford or Chevrolet stood gleaming in the downtown showroom window.

The morals of the communities in the gold mining country were no worse than most and they were probably much better than some. It was just that most people did not consider highgrading a sin. Good church members indulged in it.

In 1937, an old church in one of the mining towns burned to the ground. A special committee took it upon themselves to open the cornerstone of the ancient house of worship. They did this without the public, and especially without the press, being present. They knew that when the original church was built during the 1870's, it had been the custom to put mementos and articles of value in the cornerstone. They did this, as this was before the time of fire insurance coverage. If the church did burn at some future date, the congregation would have a "nest egg" with which to start a building fund. The 1937 committee did not want to be embarrassed by the amount and the kinds of valuables put there by their grandfathers.

What they actually got out of the depository is still a carefully guarded secret. However, one look at the size and the accouterments of the new church would lead one to suspect that the "insurance" fund was quite sizeable.

Once the automobile became a "must" for even the lowly mucker, highgrading became a more serious problem. To combat the practice the mines began to hire outside detectives. They, for the most part, were known as "security police." Their usual method of operation was to watch for too free spending by the mine workers on "wine, women, and song," and especially on the purchase of fast cars. The miner who indulged in any of these "sins" was watched very carefully. If caught, the mine had very little chance of getting any of the gold back. The highgrader was merely fired and blacklisted.

Being put on the list of known highgraders was serious, but it was not "fatal" for the experienced miner. There was always a demand for those with skills as miners in the "low grade" mines, as well as in the copper, lead, and silver mines where highgrading was not possible. Even during the worst part of the Depression, good miners were always in demand.

Being blacklisted did, however, mean that the miner had to "tear up stakes" and move away from the almost perfect climate of the Sierra Foothills. From that point on, they would have to work and live in some desert mining camp, where life was raw indeed compared to the comforts of the California towns. The unlucky miner was also cut off from more easy extra money. From then on, he would have to work overtime to buy that new car.

The security police employed by the mines were, of course, known to every resident of the mining town. Mostly they were shunned as if they had hands and fingers being eaten away by leprosy! The first security man's family was shunned by the townspeople, and especially by the children of the community. There were a few notable exceptions. In one case, the daughter of a notorious highgrader was told to make friends with the security officer's daughter of about the same age! She did so at the risk of being avoided by her former school friends. (She was!)

In time undercover agents were also employed. These were men with some mining experience brought in from the outside to watch and report any excessive appropriation of gold by either the miners or the mill men. Mill men were not too hard to find to do the undercover work. But to get experienced underground men who would inform on their co-workers was next to impossible.

In the first place, there was a "code" which

forbade the passing on of this kind of information to the management. In addition, it was not too healthy to be identified as a "stool pigeon" by the underground workers. The hazards of deep rock mining were great enough without having fellow workers looking for a situation where a carelessly balanced rock could come tumbling out of a stope at just the right time, or a missed count of the charges during blasting might let the unsuspecting mucker drive his pick or his shovel into an unexploded stick of dynamite. A missed count in the mine could be much more disastrous than a missed sponge in an operating room!

One of the first of the mine detectives was James Moyles. He was retired from the San Francisco police force. He was employed by the Empire Mine in 1919. He installed the system whereby the underground men were forced to leave their street clothes in a "change room" when going on shift. When they came off shift, they were forced to take off all their underground work clothing which they left in a dry room to be picked up at the beginning of their next shift. They then passed through a shower room, where in addition to getting rid of the dirt and grime from the working areas, they were exposed to strong ultra-violet light to compensate for some of the sunlight they had missed. Only then could they get back to the street clothes they had removed when going on shift.

In addition to the stopping of the carrying out of gold in the pockets of the miners, Moyles also began the inspection of the miners' lunch buckets. When getting off the "man skip" at the collar of the mine shaft, the miners were forced to put their lunch buckets on a table where they could be inspected while the owners were showering and changing. All this was, of course, a real challenge to the miner who had seen the highgrade come tumbling down off the face of a stope.

Any number of methods were worked out by the miners to circumvent this inspection. Almost always there was a confederate working with the underground man, usually some one working on the surface, but not necessarily so. For example, two miners working on the same shift, but in different locations in the mine, had lunch buckets which were exactly alike, even to the extent of having the same initials scratched on the outside. Miner number one came off shift with an empty lunch bucket. He placed it on the table where it was inspected and passed. He then hurried through his shower and stood waiting until miner number two walked away from the up-skip and placed his gold ladened bucket on the table to be inspected. Immediately, miner number one walked over to the table and casually picked up the "twin" of his own bucket, walked away chatting with some friend about the soccer game to be played that evening. His own bucket was left for a second inspection.

The Idaho-Maryland Mine later stopped this juggling of buckets by forcing the miners coming off shift to give their bucket directly to an inspector, and after showering and dressing they picked their bucket directly off the end of a conveyer belt which could only be reached by the miner coming out of the change room.

The mine surveyor was often the favorite confederate of the underground man. All the underground workings had to be carefully mapped, so a civil engineer was constantly going underground with his instruments. Upon the completion of a day's work, the surveyor would unscrew the lenses from his transit and fill the hollow barrel with nearly pure gold which the miner had hidden in a prearranged spot. He then replaced the lenses, packed the transit in its leather case, and nonchalantly rode the skip to the collar of the mine shaft and drove away in his car. An old fashioned transit with its twelve inch barrel could hold nearly fifty ounces of gold if it had been carefully separated from the quartz and then packed tightly in the hiding place.

Another ruse worked well in the smaller mines where automatic machinery had not yet replaced hand work. An ore car or a skip loaded with mill rock to be dumped into an ore bin would carry a distinguishing mark of some sort. Seeing this mark, the "trip man" knew that he was to pick a smoking-tobacco tin from just six inches under the surface in the front right hand corner. After all, the metal would foul the stamp mill, so he was just doing his duty by picking out the scrap iron. At least that would be his story if he was caught.

Still another method was to send up a whole car or a skip load of waste rock. In the center of the supposedly waste material would be a large bunch of highgrade. The exact location on the waste dump where this particular car load of waste was dumped would be carefully marked. Most of the highgrade could be picked up later as it was not uncommon for "snipers" to work over the newly spilled waste rock on the dump of an active mine which was known to produce highgrade ore. An experienced retired miner could often add to his retirement income (if he had any) by returning to the mill, rock which contained values, but which had been overlooked in the dim light of the stopes. The mine paid him one half the value of any rock it considered rich enough to process. If he was cooperating with a highgrader he, of course, held back the highgrade that he had been tipped off was coming up.

The biggest security problem was not the highgrading from underground. Ending the big hauls was the main duty of the security men. They knew, and the mine knew, that a certain amount would always be smuggled out by one ruse or another. The mill men were of much more interest to the police. These workers always had access to large amounts of gold in almost pure form, and even a small amount of amalgam could run into a large dollar value. The temptation to take some of this was almost irresistible and the potential loss to the mine could be very great.

Because the mercury in the amalgam is highly poisonous, the mill men could not follow the example of the South African diamond miners who swallowed the evidence. It would be recovered later, after nature had taken its course. There were many other ways, however, of getting the gold and the amalgam off the mining property.

One of Moyle's first recommendations after taking over the job at the Empire Mine was to have an enclosed, elevated walkway built from the superintendent's private office across the mill yard and along the full width of the 80 stamp mill building. In this enclosed "cat walk," the superintendent or the security police could leave the office, walk across the yard, and then continue on across the mill without being detected by either the mill men or by the office workers who might be cooperating with the mill workers. Not only could the inspectors leave the office without being noticed, they could also watch from the passageway without being observed by the men working below. Since the mill men now had no idea of whether or not they were being watched, they were forced to be doubly cautious about pocketing amalgam. All kinds of schemes were worked out to get the gold off the tables or out of the battery box and out of the mill.

One mill man was especially proud of his finger nails. He kept them long like a woman's and they were wonderfully well manicured. As he walked by the mercury tables below the stamp mills he inadvertently dragged his hand along the top of the table. This he did in a manner to resemble a careless gesture. In reality, each of the carefully manicured nails scraped up a small amount of amalgam. As he walked on up the ramp to the stamp mill above, he slipped his hand into his pocket and deftly cleaned each nail with the small shingle nail he carried there for just that purpose. The small amount of amalgam under the nails would seem to be insignificant, but actually about one dollars worth of gold was scooped under each nail each time he had occasion to walk by the tables. Usually he could find an excuse to walk by the tables four or five times each day. Thus he was able to pick up from fifteen to twenty dollars in addition to his wages. This could count up in time.

When the mine started making even the mill men change their clothing when coming on and going off shift, the man with the long nails took the small ball of amalgam out of his pocket just before quitting time. This he stuck in the center of a large wad of chewing tobacco. He had taken up the "nasty" habit in anticipation of just such a ruling by the mine.

Another ruse was to fill the fuel chamber of a cigarette lighter with the amalgam. The lighter was then put in the lunch bucket to be taken home. There it passed inspection for a long time before the guards got wise.

Bob Cole and his friends in the mill never smoked anything but Bull Durham or Dukes Mixture, which meant that they had to roll their own cigarettes. No "tailor-mades" for them. Each time they rolled one, a small pinch of amalgam went into the tobacco bag and that, in turn, went into the lunch bucket at the end of the shift. Again it passed inspection for a long time. When the trick was finally discovered, they carefully rolled the amalgam into the butt end of the cigarette. The butt ends of the smoked cigarettes were salvaged by a confederate whose duty it was to clean up the trash cans.

Several cases of severe mercury poisoning put a stop to this practice even before the mine found out about it. The heat of the burning tobacco was sufficient to vaporize enough of the mercury to make the smoke lethal when the cigarette was smoked too close to the end.

Frequently small nuggets taken from the battery boxes of the stamp mills could be washed free of amalgam and swallowed. This could not be discovered without X-rays. The mine never resorted to such desperate steps. This was the one sure way of getting pure gold out of the mill or out of the assay office. No one was ever caught at it.

Had the mines operated until after extremely strong, thin plastics became available, the mines might well have been forced to use the fluoroscope on some of the workers coming off shift. There is no doubt that any number of men would have taken the chance that the plastic wrap would hold until the small packet had passed through the digestive system.

Mike McGuire was so unfortunate as to become afflicted with a rectal cancer in his later years. The surgeon was successful in removing it all, but he was forced to give Mike a colostomy. Mike had been from the "old school," he had taken a bit of highgrade when he worked underground, but never more than enough to add a few dollars to his wage just before Christmas to be sure that his family had a few needed things which he otherwise could not have afforded. During his last years he often lamented, in a joking manner, that had he had that convenient pocket in his side during his working days, he would have been one of the richest men in Grass Valley!

The introduction of the ball mill and the floatation system for the recovery of the values took most of the chances for highgrading away from the mill men. The sluice box at the Idaho-

Maryland Mine presented a temptation, but they were locked by three separate padlocks, each operated by a separate key and each key was carried by a different man, one being the mine owner. It is not likely that much of that gold was stolen. The concentrates from the floatation cells was left in open bins waiting to be shipped to the smelter. But they were not of enough value to warrant the hauling away of amounts needed to make the risk worthwhile.

Charley Mitchell is supposed to have picked up an extra couple of thousand dollars a year by shoving a few pounds of sulfides under the seat of his truck when he stopped for gas on the way to the smeltery. He drove the truck which made the twice-weekly trip from Grass Valley to Selby. On his return, he put the truck in his own garage overnight rather than driving directly back to the mine. During the night he removed the sulfides. He put them in a pit under his garage floor. In time, he had accumulated a ton or more, enough to pay to haul to a small mine in Nevada where the valuable sulfides were mixed with the concentrates of that mine and shipped to the smeltery in Salt Lake City. He was paid on the basis of the assay of the material he delivered.

The mine got suspicious after repeated discrepancies in their records and the smeltery's weights. The gross weights were the same, but the weight of the load was often different. The twenty pounds difference, however, was too small to stand up in court. There could be that much variation in the scales from time to time. Anyway the mine got a new truck driver after the shortages became chronic.

In the early days of Virginia City, Nevada, the concentrates from the mills were well worth highgrading. Some of them were nearly pure silver sulfide. Each pound contained about ten dollars in silver and there was always enough gold mixed in to add another dollar. A pocketful of this would buy several bottles of the local firewater which was sold as whiskey. Virginia City was also the only place ever found where sulfides could be profitably placer mined. They occurred naturally in the stream beds due to the erosion of massive deposits of the ore on the hillsides above. The silver sulfide was much heavier than the native volcanic rock of the western Nevada mountains. Much of it remained on the stream beds as the storms of winter washed the weathered materials off the barren slopes. Naturally, the gold in the ore bodies remained to be recovered along with the silver compounds.

The first miners in the Virginia City area in 1852, were searching for placer gold. They found enough in the stream sands and gravels to pay for their flour and beans, but as they worked their way higher on the mountain side just north and east of Carson City, they found that they were losing most of the gold because their sluice boxes were being "plugged" by a very heavy black sand. That is, the sand was so heavy that the fine gold tended to be washed across the top of the materials which accumulated behind each of the cleats in the box.

Someone finally had the foresight to send a sample of the nuisance material over the "hump" to Nevada (Nevada City, California). This sample was included along with several samples of gold ore to be assayed. The miner who sent the sample thought that the assayer might know what the black stuff was and would know how to keep it from interfering with the recovery of the gold he knew he was losing.

Herman Ott, the assayer,[1] did not read the

---

[1] Herman Ott, was a nephew of Captain John Sutter. He was a chemist who had come from

instructions correctly. He believed that this was just another sample to be tested for its gold content.

The assaying process differs with the type of ore to be tested, but basically it consists of reducing the rock to be assayed to a fine powder. This is mixed with sodium carbonate (soda ash) to reduce the melting point of the rock and to form a sufficient volume of "slag" to allow for the settling of the molten metals. Lead oxide is then added to produce metallic lead in microscopic beads which will pick up the gold as it settles to the bottom of the crucible. Finally, a known weight of pure silver is added. This is to make sure that the gold can be separated from the silver which is always alloyed with the gold in nature in varying amounts.

The mixture is then heated in a furnace until all the materials are fused and the heavy metals have settled out. The "button" of metals is then heated in an oxidizing flame until all but the gold and the silver have been burned and blown

---

[footnote 1, con't] Switzerland to California shortly after gold was discovered. He found his services in great demand once gold in quartz was found. (At first, only the trained chemist knew the secret of determining the amount of gold in a given piece of hard rock.)

The Ott assay office building in Nevada City was scheduled to have been torn down in 1964, to make way for the construction of a modern freeway. This is probably the first case in the history of highway construction where the pressure of the local people and of the historians became so great that the State Highway Engineers were forced to redraw their plans and shift the freeway fifty feet to the east in order that the historic building might be saved. In 1970, the building was made a State Historical Monument.

away. The gold and the silver are then weighed together. Then the silver is dissolved in nitric acid and the remaining gold weighed.

The silver is added at the beginning of the assay because the assayer must be sure that there is at least twice as much silver in the final "bead" as there is gold. Otherwise the acid will not dissolve the silver, i.e., no parting will take place.

Imagine Ott's surprise when the silver and gold bead turned out to be almost the size of a thimble instead of about twice the size of the head of a pin! It weighed almost twenty grams. He started cussing his helper for playing a practical joke. He assured him that it was not funny.

The helper pleaded innocence; so the following day, Ott ran another sample, this time making sure that it was not "salted." Once more he got the great bead of silver. He continued to run more assays until all the material from Nevada had been used up. Then he made his report. The troublesome black powder was almost pure silver sulfide. It was worth about ten dollars a pound plus a dollar in gold. Together they had a value of almost $21,000 per ton![2]

---

[2]James Walsh and Joseph Woodward were the miners who had taken the trail over Donner Pass and had followed Washington Ridge on down to Nevada, California, to deliver the ore samples to Ott. When they got the assay report, they were the first to race back to the mines near Virginia City. To prove that the assay was no "fluke" and that it had not been salted, they sent a mule train load of the black powder to San Francisco. After all expenses were deducted, their check was for $160,000!

Among the first of the miners to leave the

When Ott told the boys at the bar that evening about being so stupid as not to recognize the black powder for what it was and its true value, he started such a rush to Nevada that both Nevada City and Grass Valley were almost depopulated. Everyone who could find transportation or was strong enough to pack some food, blankets, and mining tools on his back joined in the rush to the barren hills just to the east of Lake Tahoe.

The result of this rush was the production of several hundred million dollars in silver and gold. This great wealth came pouring into the United States Treasury just when the nation's credit was being stretched to the utmost by the demands of the Civil War. The great flood of gold and silver from Nevada, along with the gold from California, without a doubt had a great bearing on the outcome of the War and saved the nation!

The stories of highgrading from the rich

[footnote 2, con't] relatively rich diggings still being worked on Deer Creek at Nevada City was one George Hearst. He reached the Comstock in time to stake a claim along the main lode. That stroke of luck led to the founding of the Hearst mining fortune. William Randolph Hearst, founder of the newspaper empire, was George's son. And of course, Patty, of more recent kidnapping fame, was the latest member of the family to gain world-wide recognition. George Hearst also had the foresight to take some of the Comstock money to finance the Homestead Mine at Lead, South Dakota. That mine had the distinction of being the only major gold mine in the U.S. to continue operations during all the years when the price of gold was held to the artificially low price of $35.00 per ounce by government edict. It was, and it continues to be the solid basis on which the "Hearst Family fortune" is founded.

mines of the Great Basin Region have, for the most part, been lost. Unlike the mines in California, almost without exception, the desert mines had short spectacular lives. They boomed for a time, then, just as suddenly as they had burst upon the mining scene, they vanished. Now they are marked by rotted headframes and by crumbled mill buildings. At best, they are semi-preserved "Ghost Towns," tourists' traps to titilate the imagination of the traveler or of the history buff.

That highgrading in large amounts did take place is undisputed. To have been otherwise would have denied human nature. But those involved, both the looters and the victims, have long since moved on. Their descendants either did not know of the illegal activities of their ancestors, or they preferred that the stories be buried along with the bones of the prospectors, the mine workers, and the promoters.

Death Valley Scottie, without a doubt, got his wealth, not from a secret mine hidden away in the stark barren mountains along the Nevada-California border, but from the highgrade gold buyers in the saloons of Tonopah and Goldfield, Nevada. But one story of more recent origin can be authenticated.

In the early days of the Comstock, the amount of highgrading must have been phenomenal. This must have been so because the ore was rich and the security precautions were non-existent. In some cases, the mine owners actually encouraged the stealing of their gold and silver. Strange as it may seem, men like Mackay, Flood, Fair, Benson, and a dozen others who owned footage along the great lode, wanted every worker in Virginia City to be a "big spender." The more stories which men such as Samuel Clemens, (Mark Twain) published about the wild spending by the miners, the higher rose the prices of the mining stock being sold on

Montgomery Street in San Francisco. These men were only interested in "mining" the "suckers" who believed every story of rich new strikes, and who panicked at every report that the mines had "bottomed out." The mine owners bought up the stock at near zero, then made "new strikes" which started wealth flowing out once again. When the stock had once more reached undreamed-of heights, the flow of gold and silver stopped once more, and again the stocks plunged. Meanwhile those on the inside had sold their stock at the top of the market.

During the depression years, the mines of Virginia City were reopened on a limited scale. Some small new ore bodies were discovered and worked out. Years before, the Virginia Truckee Railroad had suspended operations due to competition from trucks. Several of the old engines had been left standing in the Virginia City yards. In 1932, or 33, some "lucky" miner was able to smuggle about a half a ton of the almost pure silver sulfide out of the stope he had been working. He was not sure that he was not being watched, so he dared not try to sell his treasure while he was still employed by the mine.

He looked for a good safe place to hide his $10,000 worth of ore. By happy coincidence, his path from the mine to his boarding house led across the long abandoned railroad yard. The old engines had not been moved for many years and the chances were that they would stay right where they were until they rusted into dust.

What a perfect hiding place! The highgrader pushed his bags of highgrade well back in the fire box of one of the locomotives. There it stayed unknown and undetected until the Japanese, in their frantic preparation for the war which they knew was close at hand, bid up the price of scrap iron. Every old mine throughout the west was stripped of its old mills, steel rails, old cables, and of every other iron part which could be hauled to the waiting ships on San Francisco Bay.

Meanwhile, the highgrader who had put the ore in the old engine fire box disappeared, probably killed in an automobile accident or in a barroom brawl. His secret died with him.

The banks which still controlled the assets of the old railroad saw a chance to collect a few dollars. They sold the abandoned locomotives to a Reno salvage company. That company had their scrap iron loaded on trucks and transported to their Reno yard. There the old engines sat for another year. In the spring of 1941, they decided that the price of scrap iron could go no higher, so they began to break up the engines preparatory to shipping them to the Japanese buyers in Oakland.

Imagine their surprise, and their unexpected profit, when they broke into the fire box of old "number 99" and found 100 sacks of highgrade!

The engines were bought "as is." Even had they felt an obligation to return their "bonanza" to the rightful owners, there was no way they could be identified.

So the Reno "junkie" threw a big party, and the Japanese "threw" the old engine back at the the American boys on the Pacific Islands in the form of bullets, bombs, and artillery shells.

One more story connected with the Comstock Mines has nothing to do with highgrading, but any tale connected with that mining district and with the times when every day saw new mines discovered and new fortunes brought up from the depths in the ore cars would be incomplete without it. The story concerns Mark Twain.

In "Roughing It," Mark Twain's story of his adventures in the West, he tells of his involvement in the mining game.

It seems that he took some time off from his

employment at the Territorial Enterprise, Virginia City's leading newspaper, to join a friend and partner in a prospecting expedition into the mountains well to the south of the Comstock. They were uncommonly lucky. Their burro led them straight to a rich lode on the hillside just north of Mono Lake.

They staked their claims and returned to Carson City to record the necessary papers. That accomplished, Sam--he had not yet become known as Mark Twain--found that it was necessary for him to go to San Francisco on urgent business. The partner remained behind, supposedly using the time allotted between the filing of the claim and the beginning of the "discovery hole" to purchase the necessary tools and powder and return to the claim to begin the work.

The partners made a fatal mistake. When they hit town following their discovery, they both drank too much at the first saloon they could find. During their celebration they showed the "bar flies" some of their rich samples and bragged about how easy it was to find.

Sam left the next morning on the first stage. The partner, instead of returning to the claims, let the new found "friends" in the saloon talk him into playing a few hands of stud poker. He was plied with free drinks and allowed to win just enough to keep his attention on the game, and away from his real business of protecting his new found wealth.

When Sam returned he found that his partner had been kept in the poker game and in bed recovering from the effects of the bad whiskey he was given, until the time allowed to commence the work on the claims had expired. At the stroke of mid-day on that fateful date, he was unceremoniously tossed out in the street and held up to public scorn.

Of course the conspirators had sent an

expedition to the silver mine where, at one minute after noon on the date prescribed by law, they jumped the claim. They made doubly sure to file the new claim notices and to do the "discovery work" as set forth in the rules covering mineral claims.

When Sam returned to find that he had lost a fortune [the claim did produce several millions] he was philosophical. He shrugged his shoulders and returned to his reporting and to his typesetting, with the observation that "such is fate."

Posterity may well balance the million dollars or more in Sam Clemen's pockets against so much of America's best literature and then "Thank God, that such is fate, indeed!"

Sluicing Near Nevada City, 1852

— Photo Courtesy of California State Library

CHAPTER XXIV

CALIFORNIA'S WEALTHY MEN

With very few exceptions the great mining fortunes came from the Comstock mines in Virginia City, Nevada. In reality, although great riches were dug from the barren hills of meta-volcanic rock which made up the country rock of that area, the really rich men got their fortunes by manipulations of the stock market on Montgomery Street in San Francisco.

In California, the riches were much more widely distributed. The mines were developed without the formation of stock companies. Given that fact, men were forced to gain their wealth from digging for gold, not by mining the suckers by stock manipulations.

The fortunes of Stanford, Hopkins, Crocker, and Huntington, the notorious "Big Four" of Central Pacific Railroad fame, came from transportation and from the construction contracts for the building of the first transcontinental railroad.

Other fortunes, such as those of Miller, Beale, Baldwin, and Bidwell came from land; from the products of the land; and from the fortunate locations of their lands in the path of city development.

Billie Bourn took about five million dollars out of the Empire Mine and pyramided it into a great fortune by developing the Spring Valley Water Company. Errol MacBoyle saw the value of his controlling interest in the Idaho-Maryland Mine skyrocket to fifty million dollars or more just before World War II, only to become almost worthless before his death in 1948! But George

Starr, Fred Nobs, and Arthur Foote, three of the most famous names associated with the Nevada County mines, were only superintendents and managers of the mines. They each lived very well, and they left their families well provided for, but not so much as to cause them to worry about estate and inheritance taxes.

Probably the greatest fortune of all those having their inception in the early mining days of the West, was that of Henry Miller of the Miller-Lux Partnership. Again the fortune came not from the mining of gold or silver, but in supplying the real miners with the necessities of life, in this case beef.

Whether or not the fresh meat they sold to the miners came from the herds stolen from Sutter during the winter of 1848-49, history does not say. Judging from the operations of the pair at a later date, it does not stretch the imagination too far to associate them with the rustlers who rounded up Sutter's cattle, butchered them within sight of his ranch headquarters at the Hock Farm on the Feather River just below Marysville, then sold the meat to the miners, and in some cases, at least, to Sutter himself after they had driven him off his own land at the point of a gun.

As Miller did not arrive in California until 1850, he could not have been in on the first raids on Sutter's herds, but the nucleus of his great herds undoubtedly came from some of Sutter's stock which escaped the butchers' knives in '48 and '49.

It was some time after the first great influx of gold seekers that the partners became prominent. Perhaps, notorious would better describe their unique position in the scheme of things in California during the one hundred years which stretched from their arrival until close to the middle of the twentieth century.

Miller, the rancher, produced the beef by one means or another and began the acquisition of the great empire of land, a great block of which is intact today under the management and ownership of the Kern County Land Company. By 1885, the partners could boast that they could drive a herd of cattle from one of their ranches along the Columbia River in Oregon to the Mexican border, and corral the cattle at one of their own ranch headquarters every night during the drive. Eighty percent of the drive would be over their own land!

Some of their business practices were questionable (if not downright illegal) even under the ethics of their day. They were party to the almost outright confiscation of many of the great Mexican Haciendas. True, the United States Courts finally upheld the validity of the Mexican Land Grants, but as it was planned before the decision had been handed down, the true owners had died or had disappeared without a trace. Or years before, they had become so impoverished that for a pittance they had given up their birthrights by signing a quit-claim deed.

Millions of acres of the great land empire were acquired under the Swamp Land Act. This was due to the U.S. law which provided that any land a man could row a boat across at any time during the year could be claimed by anyone who would file on it and pay $1.25 per acre. The theory back of the enactment of this measure was that the Homestead Act of 1862, should not apply to land of this nature; it would simply cost too much to drain it to make it useful. Futhermore, one hundred and sixty acres, the amount which could be granted to one individual under the Homestead Act, was much too small to warrant a drainage project. More should be made available to anyone who would undertake such a project. This was fine in theory, and when honestly applied, it led to the development of the world's most productive land in the delta of the San Joaquin and the Sacramento Rivers. But as usual, the greedy and the unprincipled used it to gain an unfair advantage over the rest of the population.

The story is told that the partners, Miller and Lux, used the newly enacted law to get title to about a quarter of all the land in the San Joaquin Valley. Most of the land in the Delta, the land they could have acquired legally by rowing a boat across it during the spring floods, was covered by tall tules and was of no use to them. It did, however, give them an insight as to what could be accomplished on several million acres of the flat treeless land stretching from the Tehatchapi Mountains south of Bakersville to, and beyond, Stockton to the north.

At first they were "slightly honest." They did indeed row the boat across the land they wanted. They did it, however, by falling a few cottonwoods across the San Joaquin River at a strategic spot. Then they waited until the rush of the melted snow waters of the high Sierras found the usual river channel no longer available. It did what any water must do under those circumstances; it spread out across the limitless perfectly level lands on either side of the blocked channel.

Miller rowed across each quarter section on the west side of the valley. Lux pointed his boat towards the mountains in the east, pushed his boat several miles until the limits of the flood were reached; there a team and wagon waited for him. The boat was lifted into the wagon and the team whipped to a fast run for exactly one mile to the south and relaunched. Lux then rowed back to the center of the river, but instead of stopping there, he continued on to the west as far as possible. Waiting for him there was another team and wagon with the hired driver. This time the boat was hauled three miles to the south because Miller had already crossed to the eastern shore on the half mile line, and back west to cover his two sections.

The partners were big strong dedicated men. Each could row at about two miles an hour. Each could therefore cross and thus acquire title to two sections (1280 acres) per hour. Sleep was forgotten during the periods of the highest water. They rowed from eighteen to twenty hours per day. When the water subsided, they appeared at the courthouses of the counties involved, swore to the proper affidavits, paid their $1.25 per acre for the 200,000 acres they had rowed across, and thereafter pastured one pair (cow and calf) on each 10 acres of their new ranch.

The following year, when the flood waters were at their highest, the partners added a new dimension to their aquatic exercises. Near the edge of the man-formed lakes, they launched their boats in the water which had been too shallow the spring before to allow the boat to float freely. Next they hitched a four-mule team to the front of a boat. Each of the partners took his seat at the center of the boat and rowed briskly while the teamster stood near the prow of the boat driving the mules at a sharp trot across the land which was mostly dry, but which had an occasional pool of water to salve the conscience of the "land pirates."

At last, the partners got so brazen that they "docked" their boats in a wagon, sat in the rower's seat, and swung the oars back and forth while the wagon was hauled across miles of land which had not had sufficient rainfall during the preceding years to lay the dust. In this manner, they got title to most of the semi-desert lands on the west side of the San Joaquin River from Grapevine Canyon to Tracy—the land now traversed by Interstate 5 as it makes its way from the Mexican border to Canada.

Of course anyone who did not know what was going on during the acquisition of this great empire was either an imbecile, or like Grizzly Adams, had isolated himself in the high mountains where he remained out of touch with his fellow man. But that knowledge presented no problem to the partners. Miller boasted that he had on his payroll, every county official in every county where they did business. No one ever raised any question as to the legality of their titles. The payments were all nice and legal. Each official was put on the payroll of one of the Lux-Miller ranches with the title of assistant superintendent. His sole duty was to preside at the annual Christmas dinner given for the ranch hands and their families.

Henry Miller was the ranch and land expert. Lux ran the butchering and sales end of the business. At the height of their expansion, they actually controlled the price of the beef on every table west of Denver. Miller had but to hold his cattle off the market until beef on the hoof reached a price which he deemed profitable. On the other hand, if he saw a large ranch which might fit into his empire, he was able to flood the market with beef, driving down the price until the owner of the ranch he coveted was forced into bankruptcy. Naturally, Miller owned or sat on the board of directors of the bank which held the mortgage on the bankrupt property. Usually the ranch owner knew the score. He approached Miller and transferred the title without the formalities of a sheriff's sale. Almost without exception, the former owner was offered a position as the superintendent of the ranch he had just had stolen from him. Usually he accepted. The magnetism of the cattle and land baron was so great that he was able to retain the loyalty of the man he had so recently swindled!

Miller's sharp business practices extended even to his own partner. In the partnership agreement, he inserted a clause which allowed the surviving partner to buy out the interest of the heirs of the deceased partner. That was reasonable enough except that the value of the partnership's holdings was to be set by an impartial appraiser at any time chosen by the survivor within ten years after the death of the partner and was to be paid solely out of the profits of the business enterprises.

When Lux died, Miller threw the western half of the continent into a business panic. When the appraised value of the holdings approached zero, he had the appraisal made. Of course there never were any net profits. Miller kept the books. But he was most generous with the Lux children. He made them unlimited loans, the proceeds of which allowed them to live in the manner to which they had become accustomed. When the amount of the loans plus the 10% interest had reached the appraised value of their half of the properties, he called the loans. The Lux children got nothing!

Another example of Miller's sharp business practices was illustrated by his dealings with the Southern Pacific Railroad Company. He was approached by a representative of the company. It wanted a right of way over about one hundred miles of his land on the west side of the San Joaquin River. The company expected that he would drive a hard bargain. Instead, he gave them the land. He even offered to use his considerable political influence in Sacramento to see that the State issued the franchise for the building of the "west side line."

He made only one stipulation. He demanded that the Company begin the grading of the roadbed within three months and that the portion which crossed his land be completed within two years. Then he journeyed to Sacramento and applied his potent political pressure to see that the hearings for the franchise were dragged out until the grading was completed. Then he applied even more pressure to see that the franchise was denied.

The Railroad Company had spent several million dollars building what now became a high levee to keep the floods of the San Joaquin River from the farm lands he was developing. In addition, one hundred miles of good road-bed now lay across his private property.

This piece of chicanery won the applause of the people of California. In the battle of the "giants," the Railroad was already the villain. Anyone who could best the great "octopus" became a temporary hero no matter how much he too had defrauded the public in the past.

Some idea of the size of the estate left by the financial "lord of the west," can be gained by recalling that in 1963, suit was brought by the living heirs of Henry Miller against the administrators of the estate to force an accounting for some two hundred million dollars, which somehow had been misplaced! In fact, this portion of the estate, consisting of the oil royalties from the land the Kern County Land Company purchased from the estate, was not accounted for over a period of fifteen years.

The San Francisco Chronicle of January 20, 1965, carried a news item to the effect that the final disbursement of the last $40,000,000 of the Miller Estate had just been made, each heir receiving $10,000,000. This was in addition to the more than $250,000,000 that had been distributed from the estate since the death of Miller in 1918.

The heirs had lost the suit against the Administrator. The proper reservations had not been written into the sales contract when the western half of Kern County was sold. The court apparently held that the heirs made the mistake. The Administrator, being an attorney, was only acting in their behalf. He could not be held responsible for their mistakes! (Presumably, had he not been an attorney he would have gone to jail.)

Edwin Beale parlayed $17,000 which he had made as a partner of General Kearney and General Sherman in a stage and wagon train operation to the mines in 1849, into a million acre ranch in the Tehachapi Mountains with headquarters at Fort Tejon (now a Historical Monument).

"Lucky" Baldwin invested some of his poker winnings in foothill land just south and east of Pasadena. It shortly became prime subdivision property. Horse players will know the area; Santa Anita sits near the center of the tract he developed.

General, and later U.S. Senator, and a presidential candidate, John Bidwell, with the aid of Sutter, purchased a Mexican Land Grant at the junction of the Sacramento River and Chico Creek. He joined the other northern California ranchers during the "Bear Flag Revolution," and was, thereafter an influential figure in the politics of the new state.

As already told, he had his Indian peons wash out $100,000 in gold dust and nuggets from the Feather River, then he concentrated on protecting his land holdings and his livestock from the rustlers who came pouring over the mountains following the discovery of gold.

After statehood was granted to California, Bidwell was sent to Washington as a senator. While there he met and married the charming and talented Miss Anne Kennedy. The couple returned to California via Panama. From San Francisco they journeyed by river boat to the Senator's private landing on the Sacramento River. They settled on the semi-feudal estate and built "The Mansion" among the stately, thousand year old, oaks on the banks of Chico Creek. The building still stands. It is, without a doubt, one of the best examples of Victorian architecture remaining in California. For many years after the death of Mrs. Bidwell, in 1918, it served as the office and a part of the laboratories of the Homemaking Department of Chico State College. When the college became a part of the University of California state wide system, the State Department of Parks and Beaches took title to the building and its spacious grounds. Under the protection of the State, it is now open to the public.

The materials used in the construction of this gracious home were shipped around the Horn on clipper ships. The artisans for the installation and carvings of the woodwork and the placement of the other decorations were imported from Europe. Most of them remained to contribute towards the construction of the authentic Victorians found in the Mother Lode towns and in the commercial cities around San Francisco Bay.

The Bidwells, unfortunately, had no children. This childlessness was most likely the result of a most severe case of typhoid fever suffered by Mrs. Bidwell during her adolescence. The old time residents of Chico were willing to swear that the Senator was a strong virile man, well able to father children. They were wont to point to the facial resemblance to the Senator of many of the Indian children who lived in the Indian village just north of the college campus; land set aside by the Bidwells for that purpose alone. Mrs. Bidwell outlived the Senator by many years. She faithfully carried on the educational work in the Indian settlement. She also saw to it that the legacy the Senator left to the city of Chico in the form of a twenty-four hundred acre park along the banks of Chico Creek was well maintained and that its administration was properly set up. Upon her death, the residue of the estate went to the city of Chico, the Presbyterian Church, and to various charities.

The rumors of Bidwell's illegitimate Indian children were probably malicious. Both he and Mrs. Bidwell were staunch church people. From his first days in California, Bidwell led an exemplary life. Not the slightest touch of scandle was ever associated with him. (Illegitimate Indian children were not considered a scandle at the time.) Most likely, if the facial resemblances did exist, they were due to Bidwell's relatives. Brothers, uncles, and cousins flocked to Chico by the dozens once the Central Pacific Railroad was completed. Now a visit to their wealthy relative was both easy and profitable. They were all extended the courtesy of the ranch. Some stayed for months and even years, becoming more or less trusted employees in the operation of the diversified estate. Some of them were known not to have the high moral standards of the Senator.

Before the influx of the relatives, Bidwell employed many of the disappointed gold seekers. Most of these, in time, found the Indian women ready, willing, and able to supply the wants of the pioneers who had been away from white women for months and even years. For the most part, when these men moved on to other opportunities or their wives joined them in California, they deserted their Indian "housekeepers," who now were left to fend for themselves. They had no other place to go, so they took their children and moved into Bidwell's Indian village. Some of their descendents live in the village today.

Before the arrival of Mrs. Bidwell, the "Lord" of the estate, while not living with the Indian women himself, knew that it was the common practice at the time. The only record of his disapproval of the morals of his employees is a statement that he refused to allow a local citizen to occupy one of his adobe houses as long as he insisted on sharing the shelter with a notorious dance hall girl who had been banished from the

Feather River mining camps. She was chased out for slugging a miner too hard, thereby crushing his skull, when he "got wise" and refused the laudanum-spiked whiskey she was trying to induce him to drink, in order that she could more quickly relieve him of the $3,000 in nuggets in his pouch.

So even if none of the Bidwells were responsible for the light skin of many of the Indian children, the gold seekers were. The lightness of the skin and the Caucasian features caused jealous contemporaries to start the malicious rumors about the good Senator.

Long before the University of California installed the College of Agriculture, Bidwell began the importation of improved strains of grains, better fruit trees, and the best of stock from the famous wine vineyards of Europe. He built a wine making plant and made very large profits from supplying the bars and saloons of the cities.

When it became obvious that alcoholic beverages were the cause of so much suffering and distress among his fellow man, Bidwell had the character and the fortitude to become an early member of the Prohibition Party. He tore up the vines on a thousand acres of the best wine grapes in the world. He tore down the winery and poured tens of thousands of gallons of valuable wine into the Sacramento River. [One cannot help but wonder if the migrating salmon found the shallows and rapids on the way to the spawning grounds less difficult to navigate after passing through a mile or so of river water liberally spiked with sherry, port, and ruby-red Burgundy?]

The great financial sacrifice by the new member of the Prohibition Party so impressed the delegates to the convention when the first presidential candidate was to be nominated, that Bidwell had no competition. He did not win the presidency, but he did much to make the new party respectable.

The saga of the Stanford estate need not be dwelt on in this treatis, except to say that the "Farm," the five thousand acres of pasture land on the hills west of Palo Alto that was left to the University, more or less as an after thought, to add to the University grounds and to the millions of dollars for endowment, has, after all, turned out to be the major legacy of Stanford University.

The great suburban sprawl spreading south from San Francisco and north from San Jose has reached the University area. The administrators of the University assets have very wisely refused to sell any of the land, but have opened it for the building of everything from homes to shopping centers on long term leases only. Every lease carries an economic index escalator clause. The University is thus assured of a very large income in perpetuity, with a guarantee that, as prices go higher, and the cost of operating an institution of higher learning escalates, its income will not lag far behind.

CAPTIAN SUTTER"S FORT, SACRAMENTO (RESTORED)

CHAPTER XXV

THE IDAHO-MARYLAND MINE

The Idaho-Maryland Mine, unlike the Empire, had a very spotty history of producing profits for the owners and operators. The ledge system which ran in an east-west direction, rather than from north to south as did the other ledge systems in the district, was first worked on a highly productive ledge called "The Eureka." Lamarque, the French Canadian who had gained fame as the saloon keeper and gold buyer at Boston Ravine, is known to have been one of the developers of that mine.

After producing some $19,000,000 during almost twenty years of operating, the rich ledge was faulted and mining on that portion of the ledge system was discontinued. Meanwhile, several small mines had been opened farther to the east. Among these the Idaho Mine was most successful. Eventually, it was consolidated with the Maryland Mine in order to stop continual bickering over extra-lateral rights. Still much later, the Brunswick Mine was purchased and a new vertical shaft sunk still farther to the east.

The main ledge of this system is a "contact ledge." That is, it is quartz (and in this case, considerable pure calcite--$CaCo_3$) which was forced up along the unstable contact between the granodiorite and dioric porphyry which lay to the south and the unstable serpentine formation to the north.

The other main ledge systems in the Grass Valley Area, such as those of the Empire, the North Star, the Pennsylvania, and others, were "fissure" ledges. These quartz bodies had been forced towards the surface by internal pressures within the earth's crust, along cracks or weak spots in the "country rock." This accounted for the foot wall and the hanging wall of each vein being composed of the same materials.

For reasons which are not definitely known, the contact ledges in the Grass Valley area followed the world-wide rule. That is, they were more highly mineralized than the fissure ledges in the same district. The State Mineralogist's Report of 1882, states that the Idaho group was producing ore which averaged $45.00 per ton. At the same time, the Empire and the North Star were averaging only $30.00. In 1935, when much lower grade ore could be worked at a profit, the Empire rock was averaging only $8.00 per ton, while the Idaho ore was milling out $15.00.

According to Edmund Kenyon, a life-long writer for the Grass Valley Morning Union and author of The Northern Mines, at that time, the Empire was crushing 1,500 tons per day. The Idaho-Maryland was producing almost as much gold on a tonnage of 750 going through their mill. This was from the 1935 report to the State Mineralogist.

Although contact ledges are usually richer than the fissure ledges in the same mining district, they have their drawbacks. As a rule they are much more apt to be faulted and more easily lost when they are followed to a fault plane.

The history of the Idaho-Maryland Mine is replete with a succession of losing the ledge, closing down, selling to another group which might, or again might not, pick up the lost ledge during their explorations. If they found the ledge, they usually found a good pay shoot. Profits were high until the pay shoot ended or the ledge was faulted once again and lost

This process continued until after 1900, when

a major fault cut the ledge and continued search failed to turn up any evidence as to the direction the ground had moved or any indication as to the amount of movement which had taken place. At that point, the mine closed and remained closed until 1923, when the "Jock Whitney Interest" of New York got control of the mine and started operations once again. During a three year period, they took out all the low grade ore which had previously been located and which newer methods now made profitable to process. They then spent several million dollars in trying to locate the old pay shoots at still greater depths. They never were successful.

Late in 1926, just as they were about to "pull the pumps" and abandon the mine, a local mining engineer by the name of Errol MacBoyle, who had been operating the Union Hill Mine which lay between the Idaho-Maryland and the Empire Mines, bought a few shares of the "Idaho" stock. He then attended the next meeting of the Board of Directors and demanded his right to inspect the mine. After spending several days underground, and several more studying the maps of the mine, he asked for and received a lease-option on the property.

MacBoyle had seen enough to convince him that the Whitney Interests did not know enough about the local conditions to have any hope of being successful. He also was able to convince some men with money that he would be able to operate the mine more efficiently, that he could get a higher rate of recovery from the ore processed; but most of all, that he could get the ore from underground more efficiently by using the "tribute" system (pronounced "tribit" by the Cousin Jack miners). He believed that there was a profit to be made from processing more of the low grade ore which had already been blocked out. And if, on the other hand, they were able to pick up the main ledge once again, the profits could be enormous.

MacBoyle's idea worked, but it took considerably longer than he had anticipated to put the mine on a paying basis. It was not until 1927 that the main ledge was re-discovered and put into production. At that point, the profits really did begin to roll in.

Meanwhile, true to the good old American tradition that adversity must always come before success, MacBoyle had sold all the stock in his new mining company he could talk anyone into buying. The value of the stock had dropped from $1.00 to five cents a share, and no new buyers could be found even at that price. The local Chinese grocer kept the MacBoyles eating on credit. No one else would, and Dr. Carl Jones alone put up the money to keep power flowing from the P.G. & E. Company to run the compressors and the pumps keeping the deep shafts and stopes dry. He took stock at below the five cent price as his security.

The tribute workers, by spending ten to twelve hours per day in the mine, were able to make only about $5.00 per shift. Not enough good milling rock could be sent up under these conditions to meet the mine's overhead. It was about to close once again, when the tribute crew consisting of Mike McGuire, Clarence Osborne, Guy Mooers, and Jim Mariot made the big strike. Their discovery made the mine famous and produced more than $60,000,000 in gold before economic conditions forced the closing of all the mines in 1957.

Mike and his partners were old time miners. They had worked in the mine before the 1900 closing. Mike, in particular, had been working on the main ledge when it had been faulted and lost. Later, he had worked for the Whitney Interests, and had seen some things which had caused him to believe that if he could get back into the mine, and do things his own way, he could find the long lost ledge which carried the very rich ore.

Mike had seen a very faint indication of iron stain on the footwalls of the ledge he and his crew were working. He called the attention of the shift boss to the stains which might, if followed, lead to quartz which carried sulfides which, in that mine, always carried free gold. The shift boss had laughed at Mike. After all, what could an Irishman know about a mine that had been worked by the Cousin Jacks since its discovery?

Mike stole some company time and stayed on after his own shift was over. He used a sharp pick to gather about 20 pounds of the serpentine which showed the iron stains. There was no highgrade being mined at the time, so he had no trouble taking the worthless rock home. Once there, he crushed the rock and panned out the sulfides which had caused the stains due to the oxidation of the iron. He got no more than an ounce, but that was sufficient to have an assay run. As he had suspected, the sulfides carried a value of about $250.00 per ton.

Mike had spent his own time and labor and he had spent his own money having the assay made. He had called the attention of mine officials to the presence of the iron stain, he felt no obligation to inform the management of his proof of overlooked values—and of course, the tiny amount of ore available could not possibly pay for the cost of mining. But Mike knew that just such an indication often led to quartz which might be very rich. So he said nothing. He hoped that he could get back into the mine if and when the tribute system of ore removal was adopted.

When MacBoyle took over the mine, Mike organized his crew and asked for a "tribute pitch." This consisted of 100 feet along a ledge and, supposedly, could be mined upward for a distance of 100 feet.

Mike had a reputation in the mines of being "pretty smart." Several times during his mining career, he made discoveries which led to the reopening of supposedly worked-out ledges. So when Mike asked for a pitch in a certain area, Bert Crase, the underground superintendent said, "No, Mike. MacBoyle and I were talking about this pitch just yesterday. We decided to work it "day's pay." You'll have to work somewhere else."

Mike had fully expected this response. He cussed Crase and MacBoyle and all their ancestors, but he did agree to work at another level. He and his crew worked their pitch for about three months. They averaged only about five dollars per shift each. At the end of that time they asked for a new pitch.

Crase was still suspicious, he still thought that Mike knew something; so again he refused the first request. Meanwhile the mine had stoped out the ore at Mike's first choice of locations. Only $5.00 ore was found; not rich enough to pay for the cost of mining and processing.

Mike accepted the secondary choice of locations once more, and again the mine worked his first choice and got nothing. Once more Mike and Crase went through the "ritual." Mike worked a pitch where he knew that only "eating money" could be made, and again the mine worked the ore from his first choice of locations.

At last, Crase and MacBoyle decided that Mike had not been trying to defraud them by using some knowledge he had gained while working in the mine years before. This time they let him work where he wished. Mike had figured that what happened would happen. He also figured that the mine would have to work three of his earlier choices without results before they would allow him to work where he wished. This time he had gone to the spot where he had wanted to work from the first, but had not dared to ask for.

Mike and his partners hired a Cousin Jack miner to do the drilling and the blasting on the five p.m. to one a.m. shift. Then after the powder smoke had been blown out of the working areas, they came to work at eight o'clock in the morning. They did the mucking and the timbering, if any was needed, and they sent up what good ore they found to be processed.

At the new location, Mike ordered the miner to set up his machine and begin drilling directly into the footwall instead of directly overhead, as they would if they were to stope out the exposed ore. After only one round, both MacBoyle and Crase showed up and asked Mike if they were all crazy?` Everyone knew that the next ledge was two hundred feet in that direction and, anyway, it had been worked out years before.

Mike said that he knew it. "He had once worked in a rich stope there, but after all, he was furnishing the labor and buying the powder; so if he wanted to waste his time and his money, it was not too much of their business!" They went away muttering something about "the crazy Irish."

At the end of drilling for the second round, and just before the miner tamped in his dynamite charges, Mike went underground and asked how it was coming?` The Cousin Jack looked up in surprise. "Mike," he said, "I sure did 'it somethin' awfurl 'ard in 'ere. Struck it abouts a foot in. I never did gets all the way through 'em!"

Mike told him to go ahead and shoot it anyway.

Sure enough, when Mike and his partners arrived at their pitch the following morning, they were forced to muck out about four feet of broken quartz. It was just quartz however. Just white "bull" quartz. It showed no metal of any kind. They told the miner to go ahead and drill and shoot the next round.

Again Mike showed up just before blasting time. Again he inquired about the drilling. This time he was told that about four feet in, the drill had bitten into something very soft. This was a sure indication that a layer of calcite lay along the footwall of a quartz ledge, an almost sure sign that the quartz would be heavily mineralized along the contact surface. Mike went home "walking on air!" He was so sure that they had struck gold in large quantities that he stopped at the show room and ordered a new Pontiac, the first models of which had just arrived in town.

The next morning when they started mucking, they were especially careful. They examined every piece of quartz before it was sent up to the waste dump. About half way through the pile of broken rock they began to see the glint of pyrite in the dim light of their carbide head lamps; then suddenly they saw quartz liberally sprinkled with pure gold!

It turned out they did hit an eight foot ledge which carried no values except for the foot wide band which lay immediately next to the calcite deposit.

Neither Mike nor his partners had ever studied geology, but Mike, in particular, had a feel which the geologists did not have and which the books could not teach. He had reasoned that the rich main ledge of the Idaho system had been fed by the microscopic opening in the rock which carried the sulfides he sent to the assayer. Under the extreme pressures during the formation of the quartz veins, super-heated water had been forced through the tiny crack. It was this water which carried the gold in solution. When it contacted the quartz in the main vein, the gold was deposited. The Eureka, the Idaho, the Maryland, and the Brunswick Mines worked for years on that portion of the ledge system. The engineers, the geologists, and the mine owners all made one important mistake. They had thought that the pay shoot had been lost due to faulting. For years they had been following false leads. They had now realized that all the millions of gold they had taken from the mine had been "squirted" through an almost invisible crack in the serpentine rock. This tiny opening led from the real source of the gold. Mike, by being observant and by using his common sense, hit the main lode; the "Mother Lode" or the source of gold in the other veins of the "Idaho system."

Had there been more time, Mike would have followed the iron stain lead. Eventually he would have arrived at the point where the great ledge was contacted. He knew that the mine was about to close due to the nonpayment of the power bills and due to the lack of money to pay for the needed timbers and mill supplies. He took a desperate gamble. He cut straight through the footwall of the known ledge, sure that the source of the gold in the iron stained "stringer" must lie parallel to the tribute pitch he had been assigned.

Once the new ledge was found, the Idaho-Maryland Mine became famous. The tribute miners were "kicked out," and the mine began paying daily wages. [1,200 men were given employment all through the long years of the great Depression. Their wages were not high, but they were steady and, in any case, they were much better off than the millions of men who were raking leaves or selling apples on the street corners.]

Mike and his crew was allowed to work on the rich ledge until they each had $10,000.00. This took them only one month. Then they were told that their lease was up. They, of course, knew better, but there was just not a thing they could do. Their contract was only verbal and, in any case, the Mine could always find a way to get rid of them; breaking safety regulations, for example.

Years after, MacBoyle stated publicly that he had always felt that he had acted too soon. If he had known what they really had uncovered, he would have allowed them to stay on until they each had taken out enough to retire comfortably. However, the pressure from the stockholders was so great that he must get them a good return on their investment before the pay shoot ran out.

The stockholders got a good return all right. In 1928-29, the elderly barber at the Bret Harte Barber Shop regaled his customers with stories of money to be made in Idaho-Maryland stock. Before the big strike had been made he had bought 10,000 shares for five cents per share. Already it was selling for $1.00 and it could not help going much higher. In 1934, he quit barbering and began enjoying life riding around the world on palatial ocean liners. His stock was now paying five cents <u>per month</u> dividends! It was selling on Montgomery Street in San Francisco for $8.00 per share.

Dr. Carl Jones, who had continued to advance money to MacBoyle during the darkest days before the big strike, became fabulously wealthy. He performed an appendectomy on one of his patients late in December 1935. During the convalescence, Amelia Earhart was delayed in her flight from the Hawaiian Islands to Oakland to have larger motors installed on her plane. The bedside radio was giving five minute checks on her progress. Jones dropped in on his morning rounds, listened for a minute, then said, "Shaw, she should have my plane. She would have been here hours ago!"

The doctor was not just bragging. He had a Lockheed, twin motor job, a sister ship to Amelia's, except that his plane had larger engines and was capable of a cruising speed of 100 miles per hour more than hers. He did not fly himself. He kept a private pilot at $1,000 per month, an unheard of salary for that time and place; even the mine superintendents and the local branch bank managers got less. Indeed, his plane was so fast that when the Spanish Civil War broke out, the Spanish government paid him $50,000 more than the plane had cost in the first place. It was flown to Spain. There it was equipped with four 50-caliber machine guns and used to shoot down German Stukas!

Incidentally, the doctor was a charter member of the "I Hate Roosevelt Club." In the beginning, he had agreed with the president's policies, but things changed. When it became more or less an open secret that the price paid for gold by the government was to be increased, Jones had the local bank buy up the output of all the local mines. He gave the bank $100,000 in cash as advanced payment of interest on the money they paid the mines and for the storage of the gold in its vaults. The gold, of course, was all the security the bank needed.

The gold was purchased at $20.67 per Troy ounce. Within six months, it had gone to $35.00. The bank had accumulated about 150,000 ounces. A quick calculation showed the doctor that he made a gain of approximately $2,500,000!

Always before in history, when the price of gold had been tampered with by a government, the holder of any gold was allowed to profit by the increase in price (or loss, if the price paid by the treasury had been lowered). The doctor could see no reason why this should not happen again. Instead, Roosevelt nationalized gold, and called it all in before the price was increased. The doctor had a slight stroke, but managed to survive only because he dedicated himself to the anti-Roosevelt campaign.

He had "I Hate Roosevelt" buttons made and distributed free throughout the western states. He even had several three inch buttons made for himself and for a few of his special friends. He wore his at all times. Rumor said that he even wore it in the operating room. He was quoted as having said that "he would rather be seen walking down Mill Street without his pants, than to be caught out in public without the derogatory button pinned to the lapel of his coat!"

Dr. Jones, in addition to his "I Hate Roosevelt" campaign was so affected by his "loss" of $2,500,000, that he spread the most outlandish stories about the President. For example, he made no attempt to conceal the fact that he was the source of the story that Mr. Roosevelt never did have Infantile Paralysis, as the public was told. Rather, he had had a most virulent case of syphilis! This, according to Jones, had affected his brain as well as paralyzing his legs. The doctor swore that he had treated Roosevelt in Walter Reed Hospital during the time that he, Roosevelt, was Under-Secretary of the Navy. Indeed, Jones was a Navy doctor during the War, and he did treat the President-to-be, in Walter Reed Hospital. The public was told that the Secretary had a very bad case of pneumonia. But Jones told his version of the story so often that he came to believe it himself. To his dying day, Dr. Jones cursed the American Medical Association for backing the politicians' story and steadfastly refused to disclose the diagnosis he had made; the correct one.

In the doctor's behalf, it must be acknowledged that he was an "Institution" in the mining country. He would never refuse a request for aid, and many an isolated miner owed his life to the doctor making a midnight trip over impossible roads, in impossible weather, to treat the victim of a mine accident in some isolated camp. That he would never be paid made no difference. In most cases, he never bothered to send a bill for his services.

No one will ever know how many thousands of

dollars he spent over the years grubstaking miners who were down on their luck and needed eating money. Nine out of ten times not a cent was ever returned to him. His services were also free to the children of the orphanage maintained by the Catholic Church in Grass Valley. He was not a Catholic, but no one objected more than he when the Diocese moved the orphans to Sacramento in the mid 1930's. The parish priests received a bill when they called on him to help the Lord in curing their bodily afflictions, but the sisters who taught in the parochial school, and who cared for the orphans, never received a bill, neither for themselves nor any of their charges.

Other mines in addition to the Idaho-Maryland, paid fabulous dividends after the price paid for gold was almost doubled. Dr. Jones was involved in another episode which demonstrated the effect of doubling the price of a product, while the cost falls as it did during the Depression.

In 1930, almost immediately after the stock market crash of '29, Vic Smalley was a merchant on Mill Street in Grass Valley. Like so many other Mill Street merchants (some claimed all of them), he had worked in the Empire mill until the small chunks of amalgam he had smuggled out accumulated into an amount sufficient for him to buy a store in town. He then became relatively well off by his supposed speculations in the stock market. It was remarkable how he was able to buy just the right stock immediately before a big advance. The local brokers, however, never handled any of his speculations. They only sold him supposedly gilt edged securities. These he bought after he had made one of his "killings." (Actually, he had sold more of his highgraded amalgam. The stock speculation was pure fiction.) In 1930, one of his supposed investment stocks, "Magnavox," began taking "a nose dive." His local broker was calling for more margin, so he asked Dr. Jones to take a thousand shares of Empire Mine stock off his hands at $5.00 per share. To be a good neighbor, Jones took the stock, put it in his safety deposit box and proceeded to forget about it.

Three years later, immediately after the price of gold had advanced, the Empire declared a special Christmas dividend of $50.00. The stock was now selling on Montgomery Street at $250.00!

The Magnavox stock had continued to decline. Eventually the company went into bankruptcy. Vic lost everything he had invested in it. Jones, who needed money like the Rockefellers, had picked up $50,000 in cash and had a paper profit of $250,000, all on a friendly gesture of investing $5,000 in stock that payed good dividends for almost a hundred years.

After the second World War, the Idaho tried desperately to keep going, but it was a losing game. The price of gold remained fixed while the wages they were forced to pay and the cost of all the materials they must buy kept increasing. The mine even tried to cover some of its overhead by going into the lumber business. It installed a saw mill near the quartz mill in Grass Valley. This only increased their losses. The saw logs had to be hauled too far and the marketing costs of their product in an unfamiliar field was just too great.

Eventually, like the Empire, the Idaho-Maryland "pulled the pumps" and allowed the mine to fill with water. The rock crushers, the mills, the recovery machinery, the pumps, and the hoists, plus all the other equipment needed to run a deep mine were sold and shipped to Canada where they were put to work in the rich new uranium mines.

The question always arises as to the possibility of there being more gold remaining in the mines. Naturally, all that which it was profitable to take out was mined before the mines closed. For more than twenty years the mines operated without spending the money necessary for underground exploration. Under those conditions, the profitable ore was bound to give out eventually. Also, MacBoyle died and the new management had not the imagination, nor the courage, needed to take the risks to do the necessary development work.

In 1947, as an example, when some of the younger miners returned from the service, it was assumed that the mines would go back on full production. After but a short time, it was realized that the mines had no idea of meeting the wage standards of the copper mines and of the Bay Area industries. The good miners left. Thereafter, the mines had to depend on tribute miners. There were no experienced miners available to do the development work at day's pay.

Before the young miners left, the Idaho did manage to spend some money on exploration. The "grapevine" told of another major strike near the eastern edge of the New Brunswick Mine on the 5,000 foot level. Everyone "in the know" expected the mine to begin booming again. But nothing happened, except that the stock jumped to $3.00. It had fallen to a low of $2.00 per share only a month before. But after only a few days at the higher price, it started down. It never did recover.

After some time the story of why the mine did not attempt to capitalize on the big new strike was told. It seemed that the new ore was found right up against the end line of the Idaho property. The ledge extended under agricultural land. It was impossible to follow the ledge any farther.

The owner of the agricultural land demanded $2,000,000 for his mineral rights. The mine refused to gamble such a large amount on the hopes that the pay shoot would continue on and not "peter out" after only a few feet. They knew that if the rich ore gave out immediately the Mine would be bankrupt. Of course the Mine did go broke within two years. Perhaps it would have been better to have made the investment and have gone down in a last grand gesture, rather than die slowly from the "deterioration of the veins."

In 1947, however, every one was convinced that the high cost of mining could not possibly continue, to say nothing of keeping on rising as it actually did. The owners of the mines were also sure that if the cost of mining did continue, then, certainly, the price of gold would go up. At the very least, the government would let the newly mined gold be sold on the world market.

The owner of the gold under the agricultural land still has his gold. Of course, it is buried under 5,000 feet of solid rock. It would cost a minimum of a million dollars to sink a shaft to get to the quartz. It would cost that much more to reopen the New Brunswick to get at it from the old workings.

The Empire and the North Star Mines could be opened much more cheaply since they were on fissure ledges. The hard rock on either side of the ledges would not have caved in as would the serpentine rock of the contact ledges.[1]

---

[1]Once the price paid for gold passed $500.00 per Troy ounce, the Newmont Mining Company made every effort to open the Grass Valley deep mines. One thing alone keeps the company from spending three to five million dollars to get at the gold they know can now be mined at a profit. The Pacific Gas and Electric Company cannot guarantee the delivery of sufficient power during periods of maximum usage in homes and other industries to keep the water from flooding the mines.

The power company "passes the buck" right back to the mining company. "Get the [footnote, con't.] Environmentalists—especially "Jane and Tom" off our backs. Then we will be only too glad to guarantee you a constant supply of power. Until then we can do nothing.

Amalgamation Tables in Empire Mill.

Crushed ore from the five stamp mill above fed each table.

— Photo Courtesy of California State Bureau of Mines

CHAPTER XXVI

LOST GOLD MINES OF THE SIERRA

No story of the mining country would be complete without some mention of the lost mines, real or fictional. Of all those located in northern California, the Gold Lake story is the most widely known.

William Stoddard, a Scotsman with a party of six, was supposed to have been chased by hostile Piute Indians along the crest of the Sierra near the headwaters of the Feather and Yuba Rivers. They were cut down one by one from ambush as they tried to work their way down to the mines on the lower rivers. Finally Stoddard was the lone survivor. He eluded his pursuers by swimming across a small lake during the night. Exhausted, he slept under some brush on the far shore until daylight. Then just as the sun cast its first rays over the crest of the Sierra Buttes, he crawled down to the water.

As he put his face down to suck up some of the clear cool water, the sun's first rays reflected off the surface of some yellow metallic lumps below his eyes. He forgot his thirst. He even forgot for an instant the red "savages" who were thirsting for his scalp. Throwing caution to the side, he let out a "warhoop" of his own as he grabbed with both hands and came up with a double fistful of thimble sized golden nuggets from directly off the surface of the gravel which covered the lake shore at that point. [Many prospectors, including this author, have had the same experience, only the golden glint came from the surface of bright yellow pyrite—fool's gold.]

Stoddard made his way down to Oroville and

then on to Sacramento. There, on the strength of his samples and the conviction in his voice, he was able to enlist a party of twenty miners to brave the Indians and return to bring out more of the gold.

His recruitment was made easy by the story told by old Caleb Greenwood. When he saw some of Sutter's gold, "Shucks," he is supposed to have said, "That ain't nothin', why, way back in '38, I wuz trappin' beaver high up in the mountains. I had three Squaws and a dozen or so papooses, I forget just how many. When them kids got tired o' huntin' rabbits and squirrels, the'd go down to the creeks and pick up them yeller stones. They played with them thar things, and I melted them up fer bullets when I got short o' lead. Some day, I might just get up enough gumption to go back up there and pick up enough o' them pretty yeller rocks to go on the biggest drunk ever seen since the days me and Jim Bridgers and Jim Beckwourth got into drinkin' contests in the summer rendevous up on the Yellerstone!"

The mining party was followed by half a hundred others who had heard of the story, but had not been invited to go along. They wanted to be on the spot, so that they could stake claims around the edge of the lake along with Stoddard and his friends.

The lake, or what Stoddard said was what he remembered as being the lake where he had picked up the gold, was found without any difficulty, but no gold in the gravel along the shores, nor in any of the streams which emptied into it was ever found. The party then began trekking from lake to lake in the high country, searching for the elusive fortune. The "hangers on" trailed faithfully behind, sure that the main party was only trying to throw them off the track, or starve them out, before going to the correct spot.

They never did find the gold, though they

continued the search until late in the fall. The snows of the high Sierra forced them back down into the foothills.

Stoddard now became known as the greatest liar in the mines and, try as he might, he could never get another group to join in another search. He was much too afraid of the Indians to go alone.

He told his story a hundred times in every bar and store from Oroville to Coulterville, but never with any success.

One night in a Folsom bar and store, he related his story once again, giving details as to the location of the lake and showing a few of his remaining nuggets to impress the listeners with the truth of his story. His drinking partners did not take him seriously, and as far as he was concerned, nothing came of his latest try for a new expedition.

The story was carefully listened to, however, by Ah Fong, the Chinese waiter who served the table drinks in the Folsom saloon. The next day he quit his job and secretly took off for the hills. Two months later he was back in Folsom. He sold the few belongings he had stored there, then rode the stage to Sacramento and the overnight river steamer to San Francisco. He paid for his passage with pigeon egg sized nuggets, not with the gold dust he had saved from his lowly wage at the Folsom bar.

In San Francisco, he visited a marriage broker and ordered a picture bride from China. He then invested in a Grant Street store. Later he became the head-man of the Fong "Tong." As such, he became the banker and financial advisor of the members. In all, he invested millions of the Chinese miners' money in California real estate, taking title in the names of his children, who, being born in this country, were American citizens and thus immune to the discriminatory land laws of the State of California.

The lake? You can drive to it today. Look for "Gold Lake" just north of Downieville and east of Oroville. If you are in luck, you might arrive just when the trout are biting. But you need not waste your time looking for gold.

Another lost mine story was related by Mrs. Ximena Howard of Berkeley while assisting in the accumulation of data for this book. It concerned the "Lost Kaiser Mine" near Yankee Jim's, in Placer County. Mrs. Howard was the daughter of Charles F. McGlashan of Truckee, a leading figure in the development of the Donner and Tahoe Lakes areas. He was more noteworthy for his research into the history of and the camp sites of the Donner Party. He was also the author of the first and most authentic of the many books which have been written about the tragic fate of the last wagon train which attempted to cross the Sierra in the fall of 1846.

Mrs. Howard's grandfather, David Kaiser, came to California along with the first rush of gold seekers in 1849. He mined along the American River above Auburn, with considerable success. He returned to Virginia in the late fall of 1852, and tried to "sweet talk" a Miss Elizabeth Bedford into returning to California with him as his bride.

The Bedfords were rigid Quakers. Their daughter was not about to go anywhere with this wild man from the West. He was sent packing with the admonition not to come back unless, and until, he could back up his wild stories with a hundred pounds of gold dug out of the hills with his own hands.

He took a ship back to Panama, and was soon back on the American. This time he was not satisfied with the dust from the river bottom. He struck out after one of the "blue gravel channels" where a real fortune could be made quickly. Within a month, he had his mine, and a hundred pounds of gold took only a few more months to accumulate.

Again he set out for Virginia. In addition to the hundred pounds of gold in the form of dust and nuggets, he secreted under the false bottom of the carpet bag which carried his fortune a chain of peanut sized nuggets to be an engagement present for his beloved. The chain was fully 24 inches in length and weighed well over twelve ounces.

Unfortunately, during the eighteen mile hike across the Isthmus, Kaiser was slugged and robbed by some disappointed miners returning from the gold fields. His hundred pounds of gold was gone! That did not worry him too much. He still had his mine and several of his fellow passengers volunteered to write to his prospective father-in-law and vouch for the fact that he had indeed had the amount of gold specified.

At first the Bedfords were skeptical, but in the face of the sworn testimony of the passengers, they at last relented, and Elizabeth was allowed to travel to California as Mrs. Kaiser.

In recrossing the mountains of Panama, they stopped to search for the carpet bag in the dense jungle along the trail where he was robbed. Strange as it may seem, they found it. It still had the chain of golden nuggets hidden in the bottom of the bag. The highwaymen overlooked it in their hurry to escape from the scene of the crime. (Mrs. Howard's sister still has the historical piece of jewelry, a family heirloom.)

Back at Yankee Jim's, the newlyweds were shocked to find the mine they had depended on to renew their fortune had been taken over by a group of hard-bitten claim jumpers. Rather than risk a gun battle, and knowing that his chances of getting his claim back through the courts was non-existent, Kaiser shrugged his shoulders. "Oh forget it," he said, "I'll just go out and find another mine." And he did!

His second mine was even richer than the first. During the next few years, he lived very well. He built a large brick house in Placerville. He also built and operated a chain of hotels in the mining towns up and down the American and invested in many of the early California industries.

Profiting, he thought, from his earlier experiences with the claim jumpers, he never divulged the location of the second mine. Not even to his wife. Whenever he needed money, he saddled his horse and disappeared into the mountains for a few days. He was always followed by a dozen or more down and out prospectors, bent on finding the location of the mysterious mine; but he rode cross country at night, taking care to follow terrain where no tracks would be left. He always managed to shake his pursuers. On his return, he had a pouch well filled with coarse dust and nuggets. Wells Fargo advanced him from three to five thousand dollars until the returns came from the Mint in San Francisco.

Life was rather rough for the meek little Quaker girl from Virginia. She had expected a certain amount of inconveniences, but she was hardly prepared for the hell-raising, rip-roaring antics of the miners fresh out of the hills. But for the most part she was happy. Hers was one of the better, and certainly the best-run, homes in old "Hangtown." Between keeping the house immaculate and birthing and raising three daughters, each of which in turn being rated as the most beautiful girl in the Sierra, she pretty much had her own way in their social life.

She would not contenance gossip in any form, and if it started at the Ladies' Aid Society, she picked up her sewing and walked out on the girls. She had a rather unique way of disciplining David when on occasions he disappointed her with his actions. She was not at home one day when he returned from one of his trips to his mine. He searched for her without success until he heeded the advice of one of his friends. He looked in one of the better saloons in town. To his amazement he found her there, sitting at one of the card tables, knitting industriously away! Almost speechless, he demanded to know what she was doing in such a place? Her answer set him straight in a hurry. She calmly informed him that she had found that he had spent considerable time in that particular bar during the last few months. If it was all right for him to spend his time there, she could see no reason why she should not be allowed to spend some of her free time there as well!

She had only one weakness. Though a Quaker, she was from a good Virginia family. That meant that, like all the girls of her set, she learned to ride to the hounds. She dearly loved spirited horses. One of David's brothers had become wealthy in the importing business. Mostly he dealt in the products of the Near East. He imported and shipped to California a matched pair of Arabians; a mare for her and a stallion for his brother. He was sure that any fortune based on such an elusive thing as a gold mine, must in due time be lost. He suggested that they go into the legal and conservative business of breeding fine horses.

All this was fine, except that Kaiser, who had fought the Indians as he crossed the plains on his way to California in the spring of 1849; had escaped the hazards of the early mining days; had outlived the threat of the outlaws and bandits; had by some miracle dodged the malaria and yellow fever carrying mosquitoes on four separate crossings of the Isthmus; had enlisted and fought in several of the most bloody campaigns of the Civil War, was unceremoniously thrown from the stallion while he was on his way to his mine.

When he did not return as usual from his trip to pick up more nuggets, a search party was organized. They had no idea where to start searching, so it took them a full five days to find him. It would have made no difference. Apparently, a deer had jumped from a thicket by the trail. This startled the spirited mount. The rider was thrown, striking his head on a rock outcropping, breaking his neck.

The mine where he seemingly picked up nuggets without effort for a period of nearly twenty years, was never found. It was/is, somewhere within a few hours ride from Yankee Jim's in Placer county. [GOOD LUCK]

Truth is stranger than fiction: Colonel Jack Howard, Ximena's husband (now deceased), was having lunch one day in mid 1935. He was joined by a complete stranger. They struck up a conversation. Within minutes, they discovered that they had mutual interests. The stranger, Olie Oliphant, was the grandson of the pioneer Oliphant who had developed the miners' ditches along the upper American and had taken a fortune from the hydraulic mines in that area, especially from the "blue channels" above Yankee Jim's. Howard knew that his wife's mother had been born and raised there. It seemed likely that their grandfathers had been friends and business associates. A lasting friendship was established.

The Oliphant's fortune had been depleted by the Depression. Mrs. Howard's father, C.L. McGlashan, had been a public spirited man, but not particularly a money maker. His <u>History of The Donner Party</u> was his main claim to fame. The Colonel depended entirely on his army pay. So together, they had "champagne appetites on beer pocketbooks." They did, however, have a little money to invest, a most unusual thing in the mid 1930's. They pooled their money and bought the old San Francisco Mine on the upper American River. This could have been, but probably was not, the mine where David Kaiser had taken the easy-to- get gold before he was killed, then was rediscovered but not recognized, because the real rich gravel had pinched out.

They retimbered the tunnel and began looking for the "blue channel lead." They did find some paying gravel and were just ready to start large scale operations with the aid of Dole and several other Hawaiian financiers, when the Japanese struck Pearl Harbor.

They immediately lost all of their employees to the War industries. The Government stopped its work on the "Ruck-A-Chuck" Dam, which was being built to hold back the tailings from the hydraulic mines. All they could do was to pull out and hope to wait out the war. Inflation, and the fixed price of gold stopped them from ever attempting to mine again.

The following story of the lost Cooper Mine must be read as fiction. Cooper, of course, was not the name of the owner(?) of the mine. Too many descendants and relatives of the real character involved still reside in the mining country. It is not the intention of the writer to damage in any way the reputations of any of the fine families of the Sierra Foothills. Guilt by association is still practiced in the more or less isolated communities.

"THE LOST COOPER MINE"

Jerry Cooper froze into immobility between two large boulders which lay just below the sluice box that followed the contour of the canyon wall. The moon, as it broke over the eastern horizon, cast an eerie shadow as the guard clumped down the trail, his rifle slung carelessly in the crook of an elbow. Jerry was not afraid of being heard; the rush of the water and the faint rumble of the larger rocks in the gravel being washed along and over the riffles was more than enough to cover any sound he might make. But the guard had been hired for his skill as a hunter, and good hunters were trained to detect movement rather than sound. This guard had been ordered to shoot first and investigate afterwards; so Jerry copied the tactics of the game birds and remained motionless until such time as he could get back to the work at hand.

The sluice box guard chose this particular time and place to "take the weight off his feet." Jerry had some time to reflect on just why he was playing this game of cat and mouse with the company guard, here on the slope above the South Fork of the Yuba River, on this warm July night.

Cooper had served with Grant at Appomatox, then had come west to make his fortune in the mines of California. If a little adventure could be had at the same time, it might tend to relax his nerves after the four years of fighting along the Potomac and in northern Virginia. He arrived on "The Ridge," the high land between the South and the Middle Forks of the Yuba where fortunes were being washed from the Eocene gravels by the thundering hydraulic monitors, at the very height of the hydraulic mining boom. He worked for one of the large companies until he learned the rudiments of prospecting, then he took off in search of a gravel deposit from which he could expect to gain more than the low wage paid by the mine.

Jerry mined more or less unsuccessfully up and down the Yuba for two years; then he found what he had been looking for on the north wall of the South Yuba canyon just above the town of Washington. His prize was a blue gravel channel which showed all the ear marks of being rich with Eocene gold. He dug as much of the gravel as he could from the face of the canyon wall and packed it to the river below. There, with a rocker and a pan, he recovered enough gold to prove that he really had made a rich strike and to finance a trip to San Francisco where he expected to raise the money to develop the mine.

He staked his claim in the prescribed manner, then stopped in Nevada City on his way out of the mountains to file the necessary papers with the County Recorder.

In San Francisco, Jerry had no trouble raising the funds he needed to buy the pipe to tie into the hydraulic ditch above the mine, to get the lumber for the necessary sluice boxes, hire the men needed to start the monitor, build the cook shack, the bunk house, the blacksmith shop, and to do all the other things needed to get the mine into production. He was not a promoter, but he did not have to "fast talk" anyone into risking his money on a prospect in the hills. The size of his nuggets spoke for him.

The times were such that adventurous men were accustomed to taking almost foolhardy risks. They knew this well; but the profits were so great, that men with money could afford to take the risks even if only one in ten of their investments turned out a winner.

Cooper took the river boat to Marysville, loaded the newly purchased supplies on a wagon train, and started up "The Great Highway," the road which led from Marysville up the ridge between the forks of the Yuba to the summit at Henness Pass. There, he reversed his direction, and improvising a road, he worked his way down to the mine on the South Fork.

When he got there, he found his claim being operated by a gang of "cut-throats," famous for their exploits as they fanned out from "Nigger

Tent." There was little doubt that they were associated with the nefarious Madam Romargi, the dirk carrying, gun slinging, character who terrorized the road above North San Juan and Camptonville. Jerry was met by the leader of the gang. He was given his choice: he could take a bullet through his head here and now; he could withdraw peacefully and fight for his claim in the Nevada City Courthouse, but he could only expect a bullet in his back if it looked as if he was going to win; or he could sell his wagon loads of supplies for what they had cost him, plus a small profit, and go look for another claim!

He made a quick appraisal of the situation and decided that he best not press his luck. He was sure that he could find another "lead"; and in any case, he would now have money enough from the bandit's payoff to carry him through to the opening of his next strike. Also, he figured that it was always better to be a "live coward than a dead hero." He accepted the offer, took the gold; gold which he was sure had come from his own mine, paid off the men he had hired, and sent them back with the empty wagons and with enough gold to repay the money he had raised in San Francisco.

Cooper then worked up and down the steep canyon walls of both the Yuba and the Feather Rivers for three whole years. He got nothing. His stake, which he had been so sure would last until he hit it again, ran out and he was reduced to begging mountain town store keepers for a grubstake.

Meanwhile, the bandits back at Jerry's mine, made it big. No one will ever know how much was taken out, but it was enough for those who wished to turn "respectable" to build nice houses in Nevada City, to become important business men, and pillars of the local churches.

Cooper watched what had happened. At last, in desperation, he decided to do something about it. For a long time, as he had dug and washed the gravels of the benches and the bars of the Sierra, he had been devising a plan to get at least a little of the gold which he knew was rightly his.

He studied the movements of the sluice box guard below the hydraulic operation. When he had the movements well documented, he chose a dark night and crawled up to a section of the box as close as he dared to the inlet of the water and gravel being flooded out of the gravel bank. He would have preferred to have tapped the box at its beginning, for he knew that ninety percent of the gold would be trapped in the first hundred feet of the riffles. But this was much too risky; so the upper part of the box would have to come later. Right now, eating money was what mattered.

He raised himself cautiously above the edge of the box, and disregarding the pounding of the small boulders against his hand and wrist, grabbed a handful of gravel which was trapped behind the inch high riffle. He continued up and down the box grabbing handfuls until he had all that he could carry; then he took off across the hillside to the temporary camp he had set up on a creek that ran into the river just below the mine.

The next morning he panned the gravel and decided that this beat the hard work he had been doing, and if a little was good, a lot would be better. During the next two months he made several more raids, and built up nearly enough funds to carry out the plan he had been formulating since his first attempt at high-grading. He turned all the amalgam and small nuggets he had accumulated into cash, and found that with any luck one more visit to the boxes would give him what he needed.

It was on this last raid that the guard had come along and nearly surprised him in the act. Lady Luck was with him. The guard had not seen him, and once he had snuggled down between the boulders, only by stepping on him could he have been flushed out. Finally, the guard stood up, knocked the ashes out of his pipe, relieved himself perilously close to Jerry's head, then shuffled up the path towards the "Little Giant" thundering against the cliff above. Jerry breathed a sigh of relief and grabbed a few more pounds of gravel. Then, for the last time made his way back to the camp below.

After he had panned the gravel from his last raid, he made his way down river to French Corral and caught a stage for Marysville. There, he bought the three best saddle horses from the local livery stable and found a rancher who would sell him six mules that had been pack animals at one time. Then he visited a local bar where he found two ex-soldiers down on their luck and willing to take any chance as long as the pay was right, and everything for a nice "fishing trip" in the mountains was furnished.

The pack mules were loaded with bags of junk to simulate mining equipment to keep from attracting undue attention from anyone along the trails. Then they took off for a pre-arranged camp, an hour's ride from the mine.

After setting up camp, and pretending to do a little mining along the creek at the camp site, they spent several days visiting the mine as casual observers. While there, they were able to find out the clean-up schedule, the number of men at the mine, and their probable location immediately before dark each night. They struck the day before the clean-up, however, as they knew that there would be too many men, including the mine owners, at the mine during actual clean-up time.

They put on masks and struck just before dark. They wanted to have hours of darkness in which to ride as far as possible before the pursuit, which was bound to develop, could follow their tracks.

Jerry crept up on the guard, stuck a gun in his ribs, then bound and gagged him. One of his men tossed a ten pound keg of blasting powder fitted with a short fuse into the empty bunk house, then held the crowd of mine workers at gun point as they rushed out of the mess shack to see what all the commotion was about. The other man threw two short fused kegs of explosives into the working area and lined up the four men working there until Jerry arrived.

Together, they herded all their prisoners into the cook shack and tied them securely enough to insure that they would not be molested for the time needed to turn the water out of the sluice boxes.

That done, they took flat point shovels and filled sack after sack with the gravel containing the gold and amalgam from immediately behind the riffles of the upper boxes which Jerry had not dared rob before.

Each man was then astride a good horse, and each led two pack mules ladened with the stolen gold. They made good time for two hours, then they paused to bury half of the sacks of unpanned gravel. The mules were far too heavily loaded to make time on the long pull ahead.

They pushed on up the trail to the Henness Pass road and followed that route into the high country. As dawn was beginning to break, they were close to the crest of the Sierra just south of the Sierra Buttes. Here the mountains were composed of granite, most of which lay stripped bare by the glaciers of the last Ice Age. They stopped just off the road and wrapped burlap around the hoofs of their horses and pack animals to insure that their iron shoes would leave no scratch marks on the weathered granite leaving a trail which could be followed.

After two hours of keeping to the bare

ridges, they dropped down into the basin of a secluded glacier lake. There they picketed their stock in a small mountain meadow, ate some hardtack and cold beans, and slept all through the warm afternoon. An hour before dark they took off cross country again until they picked up the trail which led down into the Sierra Valley and on to Sierraville.

They skirted that town in the dark and continued on north, and back into the hills towards Portola. At daylight, they found a mountain stream with lush grasses for their animals and camped until dark.

While the others slept, Jerry took a pan and a bag of gravel down to the water and panned out about $2,500 worth of gold! They had been lucky. The mine had been paying very well, but just before the robbery the monitor had washed out a pocket of exceptionally rich gravel. The riffles in the upper box were heavy with nuggets.

Cooper's eyes fairly popped when he saw the string of nuggets in the pan. He said nothing to the others. He knew that he would be subject to a certain amount of blackmail as soon as his "friends" had spent all their share of the gold. He wanted to be able to say that he had spent all of his too. He panned still another bag of rich gravel, but carefully hid the results. He wanted his partners to believe that the gold they were to get came from the two sacks.

The next day they rode boldly into Portola. There they put up in a boarding house. Jerry paid his helpers the $1,000 he had promised each, making sure that he threw in about $100 extra to get them in a good mood, then he invited them over to the saloon to drink to their success. He took only a couple of small drinks himself, but he insisted that the others celebrate. In fact he really loaded them. As soon as they had passed out, he paid the bartender to look out after them to be sure that they did not get rolled and to get them back to their room as soon as they could walk.

Instead of going back to his own room, he saddled his horse and put the packs back on the mules. Then he rode back on the trail to the south. He pushed the pack train hard until he came to a secluded valley. There he buried the remaining bags of gravel. Then he drove the mules well back into the hills, removed the pack saddles and halters, and threw them into a pool in the creek he had been following. He smacked each mule smartly on the rump and started them galloping farther into the hills. After a short rest, he found the trail once more and followed it to the south.

When Cooper reached the top of Henness Pass, he picked up some old, well worn, mining clothes he had previously hidden, ditched his riding clothes, strapped some mining tools on his saddle, and looking like a prospector down on his luck, made his way down to the spot where they had buried half of the stolen gravel on the night of the robbery. He waited until darkness in order to escape being detected from a distance, then he dug up the treasure and buried it near a stream where it would be easy to pan later. That done, he withdrew to a secluded spot high up on the canyon wall above and watched with grim satisfaction at the frustration of his erstwhile partners, as the next morning they came riding back, the lather on their horses' flanks and the foam at their mouths indicating how hard they had been ridden in the attempt to beat Cooper back to the buried gold.

Jerry rode leisurely on down to the Sacramento Valley. He made sure to stop at several road houses and bars on the way, to buy a drink or a meal and to pay for it with a good sized nugget. When the proprietor showed surprise at the size of the payment he pretended to "clam up," but further questioning always brought forth the admission that he guessed that he could afford what he wanted and that he could relax and enjoy life for a while. He had just found a place where a man who knew how, could pick up quite a few of these, for some time at least.

The story of the strange miner flush with nuggets he wanted to spend reached Marysville well ahead of him. He had no trouble living like a millionaire even after his bag of gold ran out. His credit was good. But at last he was forced to return to the hills to pan more of his hidden gold.

Naturally, he was followed. Every barfly and every prospector in town had been secretly watching for him to leave, supposedly for his secret mine. They followed at a discreet distance. Of course, Jerry knew that they were there, so he rode over into the watershed of the Feather River and enjoyed a camping and fishing trip as he gradually worked his way to the headwaters of the Middle Fork. Then after his horse was well rested, he rode all night to escape the few followers who still lurked in the bushes. He was well aware that they would follow his tracks in daylight, but by the time they had reached the spot where he had buried the gravel to the south of Portola, he had panned out two more bags and was well on his way back to Henness Pass, where he allowed them to overtake him. They rode back to Marysville together. Jerry paid for the food and the drinks on the way.

For two years Cooper followed this routine. Living high, then visiting the hills once again. He always returned with a poke well filled with nuggets.

But all good things must come to an end. Jerry had always been super cautious in his dealings with strangers. But this one was a gorgeous girl who at first seemed to be "hard to get." His determination to make a conquest led him to make a fatal mistake. He accepted when he was invited to her room. There he joined her in a drink. Only his was heavily laced with laudanum. She had only expected to drug him sufficiently to be able to escape with his heavy poke. She poured too much of the narcotic into the glass. Jerry never did wake up!

How many other hidden and lost mines in the western states were the result of successful highgrade raids such as Jerry's, or were the proceeds of a series of successful bank robberies, or the nuggets found when the Wells Fargo strong boxes were opened after a stage stick up, there is no way of knowing.

As previously mentioned, there is a great deal of evidence to indicate that Death Valley Scotty's hidden mine in the Pamanints was entirely fictional. The real source of his gold was the highgrade buyers in the saloons of Goldfield and Tonopah, Nevada. He was such a good showman and actor that he was able to keep up the myth of his hidden mine until his death. Even after that, those who believed in the hidden mine story, now listed it as but one more of a series of lost desert mines. Of course, it was never found.

Cooper's cache of bags of gravel buried on the north canyon wall of the South Yuba River has never been reported as found. Some day, some lucky prospector will find a trace of gold in a ravine. He will follow it up stream until he reaches its source. Then, instead of the colors leading to a blue gravel channel, he will find a pocket of gravel. His disappointment will be salved to a certain extent by the size of the nuggets he will uncover. He will know, however, that he has not found a mine, because, mixed with the nuggets will be lumps of amalgam.

CHAPTER XXVII

THE ALCALDE MINE

Every mine has it own story. Sometimes the history of the mine deals with the discovery, sometimes with events during its development, or with unusual happenings while it was producing gold or other wealth. The stories of the Empire and the Idaho-Maryland Mines have been of interest because of their great productivity, and because they were so well known for the progress in the science of mining made by the men managing and working in them.

The story of a very small and relatively unknown mine is illustrated by the Alcalde. Again the mine was in the Grass Valley area. It was mentioned briefly in the discussion on the granting of mineral patents.

The Alcalde was located about two miles southwest of Gold Hill, in what was originally known as "Dead Man's Flat." The area was typical "pocket hunting" country. Not only was there gold in the gravel of the small stream which ran through the property, but there was also gold found in bunches in the clay on the hillsides. The quartz in the ledges which outcropped through the serpentine country rock of the area had been weathered by chemical action to clay which was not washed away by erosion. Where gold had been embedded in the quartz, it remained in the clay. In mining the unweathered quartz below the zone of weathering, the gold found there was mostly in pockets of very highgrade rock. Hence, when weathering did take place, the pockets still remained in the clay which succeeded the original quartz.

The main ledge was located in the days

immediately following the cleanup of the limited amount of placer gold in the streams. Someone found that by following the clay deposits to the hard rock below, the quartz ledge from which the clay (and the gold) had come could be found and mined. The ledge, although intruding into serpentine rock, was in a fissure and because of its closeness to, and the general characteristics of the quartz itself, was thought to be related to the great North Star formation.

Who discovered the ledge while pocket hunting is not known. It was not of sufficient importance to be mentioned in the newspapers of the day. Very old waste rock dumps show that various attempts were made to work the ledge in its northern extention before 1900. The size of the dumps indicate that the pockets were too few to warrant more than half hearted attempts to make a paying mine at any of the shafts.

In 1910, a man by the name of Root, the State Mineralogist at that time, began developing the ledge along its southern extention. He pushed a shaft down to a depth of 400 feet, and eventually drifted south some 600 feet. He was lucky. He cut several pay shoots of highgrade and of good milling rock.

Visitors at the mineral display of the State Department of Mining and Mineralogy, in the old Ferry Building, at the foot of Market Street in San Francisco, will be able to view many pieces of glittering highgrade ore. These pieces came from the Alcalde Mine. The fact that Mr. Root was not only a mine operator, but the State Minerologist as well, made it possible for him to sell to the state exceptional pieces of "picture rock" at a price considerably higher than that which the Mint would have paid for the gold.

Root stoped out most of the quartz above the 400 foot level. He milled the ore. The profit from that, plus the highgrade, netted him about

$250,000. Then he began sinking a winze (a small shaft below the drift from which the ledge had been worked, rather than an extention of the main shaft) near the southern end of the ledge. At the depth of only 20 feet, the winze showed that the ledge split into a hundred stringers running out in all directions, none of which seemed to carry gold.

Root decided that discretion was the better part of valor. He quit while he was ahead. He closed the mine, sold the ten stamp mill along with the recovery equipment, and allowed the mine to fill with water. That was in 1918.

The two mineral claims covering the main vein of the Alcalde had never been patented. To keep his title to the mine, Root was forced to do the assessment work each year. This he did until 1927. At that time he bought a block of stock in the ill-fated, C.C. Julian oil empire in Southern California, giving the mine as payment.

Julian knew that the Alcalde had very little, if any value, but he also knew that his stock was totally worthless; so trading nothing for a gold mine which might, or again might not, have some value, was a good deal. He sent an ex-miner from Montana to Grass Valley to act as a watchman for the machinery which Root had been unable to sell and to do the assessment work.

The ex-miner, Joe Murphy, enjoyed semi-retirement on the property until 1932. At that time an old Alaskan Sourdough, one Jake Berger, came to Grass Valley, a deed to the mine tucked in his briefcase.

Berger had been in on the original strike on the beach near Nome. Following that, he had made and lost a half dozen fortunes. His latest fortune being in the form of a charming and wealthy widow from San Francisco. The widow's first name was Mattie, the last name does not matter.

Berger was also a confidant of Julian; so when the oil man wired from Oklahoma City that he desperately needed $5,000 to get out of the country ahead of the FBI, Berger sent the money. He accepted the deed to the Alcalde Mine as security for the "loan."

Meanwhile, Joe Murphy, the watchman, had become a very good friend of Mike McGuire, the same McGuire that made the fabulous strike in the Idaho-Maryland Mine some six years before. Mike was made to order for Berger. Here was a practical mining man, one raised in Grass Valley, and one who had been working in the Alcalde when Root had closed it down. Better still, he had a reputation for having "the luck of the Irish" when searching for lost veins. Berger hired him as superintendent after only one interview, and began operations at Dead Man's Flat.

That these operations consisted of one part mining and nine parts stock promotion, is not a part of this story; so we can forget Berger for the present, except to note, that in deference to his new wife, he renamed the Mine. Thereafter, it was known as the "Mattie Mine."

The Root shaft had caved in down to the 150 foot level. Mike hoisted out the mud, the rubble rock, and the rotted timbers. That done, he found that the crumbling serpentine made retimbering the old shaft impractical; so he started a new shaft 600 feet to the south, and pitched it to intercept the old mine drifts at a depth of 400 feet.

At 380 feet, the Finnish miners who had drifted into the Grass Valley from the copper mines of Michigan, came "top side" at 3:30 in the morning. They called Mike out of bed with the startling news that they had hit the vein, and that it looked like it was "good mill rock"!

Mike dressed and drove directly to the mine. Sure enough, the bottom of the shaft was dead center on an 18 inch ledge of quartz which now and then showed flecks of gold. About twenty tons of this ore was taken out during the sinking of the shaft the remaining distance to the drift below. The ore, which assayed $20.00 per ton, was piled just east of the "head frame." It is still there, but it has been picked over so thoroughly that very few colors are to be found anymore.

After the shaft hit the old drift, a few shots in the stope 50 feet to the north threw down "picture rock"! Why Root stopped mining and left this part of the ledge intact, is one of the mysteries of mining.[1]

When the pay ore was found, some 2,000 tons were blocked out. It averaged more than one and one-half ounces, or $52.00 per ton. With this as bait, stock suddenly became easy to sell. Some real capital was attracted for active development work. The ore was left in place. It was worth much more for stock selling than it would bring at the Mint.

The original drift was extended to the south, but no more paying ore was found before the ledge pinched off. A cross cut was driven 2,000 feet to the east in an attempt to cut the Normandie ledge at the 400 foot level. It had been lost at 250

---

[1]The writer once personally picked some rock out of a powder box in the mine office. When broken apart, most of the pieces were wired together with strings of gold. One piece about the size of a man's double fists checked out to contain almost ten ounces of gold—$350.00, at that time. Berger took that piece to New York, and using it to cast a hypnotic spell, sold some $25,000 of stock in the mine. Of course, he had to give the specimen to the New York buyer as a memento of his great good fortune in being able to get in on the ground floor of such a deal. (It was a good thing that he kept it, because that was all he ever got from his investment.)

feet. Again no results, so the shaft was sunk another 400 feet. At that point, a station was cut and a drift started in a northerly direction to pick up the original Alcalde ledge if it did form again.

Less than a week before they were to give up the search and salvage what they could from the ore blocked out above, they hit a four foot vein running parallel to the original Alcalde ledge. Unfortunately it contained only pure white quartz. Not even a sulfide was to be seen.

The absence of gold in this area is no criteria as to the value of a ledge. The original ledge of the Eureka Mine, the one which later became part of the Idaho-Maryland, was followed for 600 feet before a pay shoot was found. Once discovered, however, the shoot continued on for 6,000 feet and $19,000,000 was recovered before the ledge was lost. The closest mine to the Mattie was the North Star. That mine eventually turned out more than $100,000,000, after it had passed through a barren zone which exactly matched that which they had just encountered below the workings of the old Alcalde Mine.

With renewed enthusiasm, they began drifting along the big new ledge. After only fifty feet, great chunks of sulfides were thrown down by the blasting. In Grass Valley, that is almost positive evidence that pay rock will follow very closely behind.

At this point, money began to talk loud and clear! When the 2,000 tons of ore on the original ledge was blocked out, Berger succeeded in selling $150,000 worth of the stock to C.L. Best, a vice president of the Caterpillar Tractor Company. Best's father had organized the original C.L. Best Tractor Company of San Leandro, California. He had been so successful that the Caterpillar Company had finally merged with the Best Company to get rid of the competition.

Best had a nephew who had just graduated from the University of California as a mining engineer, so he had a double motive for getting control of the Mattie Mine. The nephew came to Grass Valley to learn the practical side of mining under Mike McGuire. But when the sulfides in the ledge at the 800 foot level indicated that a big strike was about to be made, Best put his nephew in full charge. They closed the mine, and demanded a re-organization of the company. This was done, and the first act of the new board of directors was to demand a large assessment against each share of the converted stock. None of the old stock holders could meet the demand. Best now owned the mine in its entirety.

Best's double cross of the other share holders backfired. He was the legal owner of the mine, but in reality, he had nothing, or even less than nothing. It seems that in transferring the title to the machinery and to the active claims from the Mattie Mining Company to Best, the Title Company found that Berger never did have clear title in the first place. Julian had sold some $250,000 in bonds, giving the mine as security. These bonds were still outstanding. They represented a first claim on all the physical assets of the mine. Julian had just forgotten to tell Berger about these when he had signed a quit-claim deed just prior to his suicide in Hong Kong.

Best spent two years and $25,000 in a vain attempt to find those bonds. No trace was found. He did not dare operate the mine until those obligations were accounted for. If he did, and sent any gold to the Mint, he would become liable for all the accumulated interest on the bonds, and for their maturity value as well.

Best closed the mine completely. He took all the movable machinery to the Ruby Mine in Sierra County, along with his nephew to act as superintendent. The Ruby had been a rich quartz mine that followed a twelve inch quartz ledge, north into the canyon wall of the Middle Yuba River. It was closed when World War II took all the miners to the shipyards.

After the war, the mine was reopened, but the drifters had followed the ledge less than a hundred feet when it broke out of the hard country rock into soft volcanic ash. The ledge was lost. But Best was in a mood to gamble. He continued to drive the drift straight ahead. He hoped that when the opposite wall of the canyon the ancient river had cut was reached, the ledge would be found.

He did better than that. Within one hundred feet, the drift came to the center of the canyon. A long gravel bar was deposited by an Eocene stream, then covered by the ash of an equally ancient volcano.

Hulesdunk, the superintendent, had the good sense to take out a cubic yard of the newly found gravel. He ran it through a sluice box and panned the concentrates. He recovered more than forty dollars worth of coarse gold and nuggets!

So, the old Ruby Quartz Mine became the Ruby Gravel Mine. The results were spectacular. Best made more millions to add to his assets. But before he got things under complete control following the rich strike, the highgraders had a field-day. One lucky crew member slipped a nugget into his pocket. It was as large as a big man's fist. It weighed slightly more than 100 ounces! The gold it contained was valued at $3,500.00, but of course, it brought a premium as a paper weight for the desk of a war profiteer.[2]

---

[2]The writer saw that nugget as it was being transported to San Francisco to be offered to the highest bidder. It brought $5,500.00. The 1987 value is in excess of $55,000.00 for its gold content; as a paper weight, double that amount.

What happened to the quarter million dollars in bonds which Best could not locate? They were found in the portfolio of a Reno casino owner when the notorious Riverside Hotel robbery in Reno was solved in 1948. However, they were worthless.

During the War, it was not necessary to do the assessment work on unpatented mineral claims. On, or before, noon on July first, 1948, work needed to be started on all such claims. Mike McGuire arrived at 11:30. When Best nor none of his men showed up to go to work, Mike drove the stakes and began the discovery shaft on the main claim. Later, he was able to patent all the worthwhile property. It became a part of his estate.

Pelton Wheel. 500 HP. North Star Mine

CHAPTER XXVIII

PHOENIX GOLD

Highgrading began with the taking of rich ores from the mines. Soon the mine workers found that amalgam from the mill or small buttons of gold that remained in a crucible after a pour, were much easier to smuggle from the mining property.

Gold is quoted in dollars per Troy ounce. Roughly $35.00, at the time the mines closed.

One ounce of gold occupies about 1.5 cubic centimeters of space.

Small wonder then that highgraders preferred bullion or amalgam to add to their take home pay.

For the older miners, the ones with families, taking a little was almost a must. Otherwise they were forced to "moonlight."

Usually, this was limited to a little farming. They cultivated one to a dozen acres before and after going on shift. A few chickens, a milk cow, and a sow for two broods of piglets per year, helped feed the family.

Some planted garden crops for sale, but most town dwellers had yards large enough to grow the common vegetables. Or, if they did need to buy extra potatoes or cabbage, the Chinese gardeners filled their wants at a price the miners could not meet.

It has long been recognized that sailors and miners were the two laboring groups which looked forward most eagerly to retiring on small farms. As a rule, they only dreamed of the day they could become their own bosses and move to homes in the country. Mostly, they failed to accumulate the money needed to purchase their dream farms and again, most who did, found that they lost money each year. Only those with capital to pay cash for acreage to operate on a commercial scale could retire to the life of farmer or rancher.

Lester Brown, a miner working at the Idaho-Maryland Mine in Grass Valley, tried valiantly to suppress his cough as he waited in Dr. Carl's office. His attempt, fooled the doctor not at all. One look at Brown's face, and the tell-tale sounds coming through his stethoscope told him that this was yet another case of "silicosis"—the dreaded "miner's consumption."

"Well," the doctor said in as cheerful a voice as he could muster, "No more underground work for you. But if you live right, take care of yourself, and don't get wet when you go deer hunting, you might be able to lick the bugs."

Brown grimaced. "I knew what you were going to say. I've seen too many men start to cough like this, not to know what ails me. But I'm going to lick this thing. Bill Jenkins did. And I'm just as good a man as he is."

"You think them experts over at the sanatorium at Weimar can do any good? Get me back on the job any quicker?"

Dr. Carl nodded. "You got the guts to follow their advice and do exactly what they tell you, you'll probably cut the time in half."

"But, I'm warning you. You'll have to get a clean bill of health from them. With what I just heard, and with what I know, I'll never sign a clearance for you to work in any mine. It'll kill you in time. Get yourself a job on the surface where you can be in the sunlight. I'll never sign for you to go underground. Not a year from now. Not ever."

Again Brown grimaced. "OK, Doc. But do one little thing for me. Call MacBoyle at the mine. Give him your report on my condition. Then tell him I'm going over to Weimar and when I get back I want a topside job. Anything just to keep working and making a little money. Tell him the mine owes me that much."

The specialists at Weimar were delighted with patient Brown. Not once did he complain at their rigid regime. He arrived in the late spring just as the sun gained height enough to insure that it's beneficial rays would penetrate the atmosphere and do so much for those who had been shut away in the darkness of the mines, and the right amount of exercise, the clean crisp air of the pine-clad hills did wonders to destroy the bacilli that were so destructive to the health of miners the world over.

Late in the fall, Brown was back in Grass Valley demanding the job Dr. Carl had arranged at the Idaho-Maryland Mine. His work consisted in keeping the mine yard neat and clean. Officially, he was listed as a gardener. Someone must have had a perverted sense of humor when that category was listed on his employment card. The nearest things to a living plant growing in, or on, the five acres of serpentine rock and iron stained soil which made up the mine yard, were a few stunted, red-barked, manzanita bushes, augmented by a number of poison oak plants, whose leaves now showed the red of their autumn phase.

On the very edge of the mine building site, two scraggly digger pines defied the sterility of the soil. Their gray-green needles stood in stark contrast to the dark greens of the ponderosa pines, the Douglas firs, and the incense cedars that luxuriated on the adjoining hills. There, granitic rocks were buried under a mantle of semi-compacted ash, blown from a very ancient volcano fifty miles to the east.

Brown's main duties were to make sure that the crushed rock walks connecting the various buildings were raked each day; to pick up any litter outside the buildings; and of more importance, to pick up and pile the trimmings, the bark, and the shavings from the open-air carpenter shop where mine timbers were being prepared to be put to a multitude of uses in the underground working areas.

The waste lumber was to be cut into stove lengths and piled near the office building, awaiting the colder months. The truly waste materials were stacked and mine employees were urged to haul them home for use in their heating and cooking stoves. Mostly, the offer of free wood was ignored. "Down" timber was so plentiful within a few miles of the mine, that the workers preferred to go into the National Forest to bring home the better fuel.

Brown was told to bag the unwanted material. Later, it would be hauled to an old gravel mine site where it would gradually weather and rot away. Wet, and rotting mine timbers and old minature underground railroad ties were also disposed of in the old gravel pit.

The assigned work did not keep Brown busy. Almost every day, he found time to go into the carpenter shop to give a hand to the workers there. He helped carry heavy timbers to the saws and planers. He stacked the cut lumber and helped deliver it to the main shaft to be lowered underground.

MacBoyle, the mine owner, noted that Brown was doing more than his assigned work. He stopped to inquire, "Why?"

Brown smiled at being recognized by the "big boss." "Well, Mr. MacBoyle," he said, "I'm not one to sit around and do nothing. Anyway, I'm building up my strength since my bout with the 'bug'. Some day, I'm going to come into your office with a clearance to go underground again."

MacBoyle patted him on the back. "That's fine. You'll always have a job here. But underground? Some have made it for a time, but don't get your hopes too high."

Two months passed. Brown continued to work hard. Then he saw MacBoyle walking to his car.

"Mr. MacBoyle? Could I speak to you for just a minute?"

"OK. But I'm not letting you go underground."

Brown shook his head. "No, Sir. It's not that. I've been wondering about all those wet timbers and ties we have to haul to the dump. I've got a big back yard. Would you let me bring my pickup, and let me take a load home with me? Given time, they'll dry out. I can saw them up for fire wood. May take all summer. But that's better than driving ten miles to find a place to cut."

Again MacBoyle patted him on the back. "Fine, fine. Take all you can use."

Brown thanked him. To himself, he said, "Step number one accomplished."

Step number two in Brown's campaign to finance the purchase of a ranch in the lower foothills was one he thought of during his confinement to his bed at the Weimar sanatarium. He had maneuvered his way into a position to get the gold off the mining property. Now he must find a way to get his hands on raw gold or amalgam. To do that, he must have a confederate who worked in the mill.

Bill Osborn, an experienced mill man familiar with the operation of flotation cells, found himself unemployed when the Depression closed the copper mines in Butte, Montana. He heard through the miner's grapevine, that the gold mines in California were, at long last, doing away with their stamps and installing ball mills. Experienced mill men would be needed.

He loaded his family into his Model "A" Ford, tied their most important possessions to the running boards and onto the roof of the little car, and camped out the five nights on the way to the booming mines in northern California.

The factory man who had installed the flotation cells at the Idaho-Maryland Mine in Grass Valley taught the mill men all he could about the operation of the new and rather complicated machines. MacBoyle was delighted when a man with years of experience walked into his office. He was put to work immediately.

On warm summer days, Osborn made a practice of seeking a shady spot on the north side of the mill building to eat his lunch. On winter days, he migrated to the southern exposure to find the warming rays of the noon-day sun.

Lester Brown was looking for just such a man. He arranged his lunch break to coincide with that of Osborn's. Soon the two men were telling their life histories. That led to fishing trips, and to many hours sitting in blinds, waiting for a legal buck to pass by on its migration from the high country to the snowless hills near the Sacramento Valley.

Brown's favorite pastime when on these trips with Osborn was to regale him with stories of the highgraders. Osborn, of course, knew of such activities, but in the copper mines of Butte, the ore did not warrant the risk of stealing.

Once a firm friendship was built up, Brown asked if there was any way his new friend could get into the sluice box below the ball mill?

Osborn only laughed. "Sure I can get in the box. I worked for a locksmith when I got out of high school. But I wanted to see the world, so I gave that up to go into the mines. When the copper mines closed, I came here. What you got in mind? I might get some of the gold, but there's no way I can get it out of the mill."

Brown took the real risk of losing his job, but outlined his proposition.

"You get the gold out of the box. I'll get it out of the mill, and into the hands of a highgrade buyer. We split fifty-fifty. How about it?"

Osborn's first reaction was: "What the hell? You undercover? Trying to get me fired and run out of town?"

Brown shook his head and laughed. "God damn it! I ought to punch you right in the nose for being so suspicious. I thought you were a better friend than that."

He considered for a moment, trying to determine if Osborn could be trusted. Then he took the added risk. "No. I told you about my lung condition. I got to get ahead enough to buy a cattle ranch where I can work outside the rest of my life. I figure we make a deal. I get my ranch. You get enough to go into business somewhere."

"Yeah? What happens if we get caught?"

"Well. Life is full of risks. But I've got a scheme. One that's never been tried before. We work for about a year. Keep sober. Keep our mouths shut. Not much chance of anybody finding out. We stay friendly, but not too friendly. Just enough to allow me to see you in the mill building a couple of times a month. That's all."

Osborn considered for a minute. "OK, I'll try it with one ounce. That works out all right, I'll go 'big time' with you. Don't tell me how you're going to do your part. Best I don't know."

"Step number two completed," Brown said to himself as he walked away.

Brown took Friday off. He drove over to Weimar to see one of the doctors with whom he had become friendly. The doctor listened to his breathing, then remarked about the completeness of the cure.

That was what Brown wanted to hear. He asked for, and got a clearance to go underground on a limited basis.

MacBoyle frowned when he was handed the

document. Brown was prepared for a protest.

"Mr. MacBoyle, I don't want to go into the drifts or into the stopes as a steady thing. Put me on one of the timber crews. You know I've been helping out in the carpentry shop. I've learned enough to qualify as a helper. A lot of my work will be topside anyway. I'll be all right."

"So that's why you were so willing to do all that extra work?" MacBoyle laughed. "OK, you deserve a break. I'll tell Crase to put you on the day timber crew. But you watch your health. You know, of course, you'll have to sign a disclaimer. Then you can't hold the mine responsible if you get sick again."

Brown thanked him. He walked away whistling the latest Crosby tune. "Three down, and the big one to go," he said.

Meanwhile Osborn managed to find a flotation cell out of order. That made it necessary for him to work overtime. He repaired the cell in five minutes, then spent two hours waiting for all the other workers to take a coffee break away from the ball mill and the sluice box below. There, with expert fingers and with the tools from his apprentice days, he found the "action" of the tumblers in each of the locks. Knowing that, the making of a set of keys was elementary.

A few days later, he tapped the box and removed several ounces of coarse gold. He secreted his treasure under one of the flotation cells and waited for Brown to claim the ounce he had promised.

Brown did not rush. He worked hard at his new job. Then one morning before going underground, he walked into the mill building. He was already dressed in his underground working clothes, clothing which must be left in the change room before he could reclaim his street clothes at the end of his shift.

He walked away from the mill straight to the

down skip in the main shaft of the mine. Men coming out of the mill building were always suspect. But a worker going directly underground? That's where the highgrade came from. If the miners couldn't get it out, why take more back to its source?

"Step number four completed." Brown said as he "hefted" the weight of the gold in his pocket.

When MacBoyle allowed him to go into the mine once more, Brown continued to drive his pickup truck into the mine yard. Each day or when there were planks, stulls, or railroad ties available, he tossed them into his truck and drove through the mine gate. The guard had orders to let him through.

That was step number five. Now the real, and final step was put into operation.

Underground, Brown waited until he was working alone. From his toolbox, he took a brace fitted with an inch and one-half bit. He bored into the end of a soggy tie. Then he carefully pushed his little bag of gold into the empty space. He already had a plug made from a similar tie, cut to fit. He drove the plug home and covered the battered end with mud from the floor of the drift. He made a distinguishing notch on the opposite end of the tie so that he would know which piece of waste wood carried the gold.

A day later, he loaded the scrap material from the drifts and stopes onto the up-skip. That evening the pile fitted nicely into the box of his truck. Within the hour, the wet and soggy wood was piled in his back yard to be exposed to the drying breezes of the Sierra foothills.

Two months later, Brown cut sixteen inches off the end of the marked tie. He split it to fit in the firebox of the kitchen stove. It was still damp, but dry enough to be ignited by the other firewood. When all had turned to ashes, he swept everything remaining in the stove into a gold pan.

He had an old wash tub under the down spout of his house. It was made to order for the panning of the ashes, leaving a string of bright yellow metal in the crease of the pan.

Next he saw an illegal gold buyer and traded the ounce of gold for $30.00.

A few days later he invited Osborn for a ride to Sacramento to see a show. On the way, the cash from the buyer was divided.

"OK," he asked. "You ready to go to work?"

By this time Osborn was drooling. "Sure am. Can you handle 25 ounces?"

Brown was too excited to answer. He only nodded, then added: "All you can get. But be careful. Don't take too much at first. They might get suspicious."

"OK," Osborn said, "I'll forget my lunch bucket Monday morning. You be a good friend. Bring it to me as soon as you get ready to go underground. I've been taking almost ten ounces every day. Got enough to keep you busy for quite a while."

When the cold weather of winter arrived, Brown did not split the marked timbers. He cut them into fireplace lengths. Periodically he swept the ashes off the fireplace floor to be panned. For a whole year he continually referred to the loot as their "Phoenix Gold."

Osborn never could figure out what he was talking about.

At the end of the first year of the operations of Osborn and Brown, each had the "stake" that they had aimed for. Brown began looking for a good cattle ranch which might be for sale. The Great Depression still held the world in its vice-like grip. Almost anything could be bought for salvage prices. Land, in particular, was nearly worthless.

Grass-fed cattle were selling for five cents a pound, hardly enough to pay the trucker to carry the beef to market. Seemingly, anyone with money would be mentally unstable to pay even the going price of $25.00 per acre for winter range land on the first swelling of the foothills.

But there were always exceptions.

Old Bill LeDuc's grandfather had hit a big pocket in 1860—or at least that was the story. The grandfather, Bill's father, and he himself, lived quite comfortably on the 500 acres lying to the west of Indian Springs and extending to the flatlands in the valley below. The land carried some water rights as well as grazing privileges for a hundred "pair" (cow and calf) in the Tahoe National Forest.

But now LeDuc could no longer ride. He wanted to retire to the ancestoral Victorian home in Nevada City. With the ranch buildings, the water rights and irrigation ditches, and the Forest summer grazing rights, together with two hundred head of white-faced cattle, the asking price of $50,000 was well within reason. But no one had, or could borrow the fifty thousand. The real estate brokers even refused to list the property.

However, LeDuc knew much more than did the brokers. He visited the poolhalls and the saloons of Nevada County. Without any apparent effort to sell, he let it be known that if anyone had enough pre-Depression $20.00 gold pieces buried in his back yard, or pushed to the rear of his safety deposit box, they could strike a deal.

The rumor of the ranch for sale reached Brown. He drove down to Indian Springs.

Bill LeDuc met him at the ranch house gate.

"Hi, Lester. What you doing down in this sun baked part of the world?"

They shook hands. Brown asked if there was any place they could talk without being overheard?

"Sure. Come on in the tack shop."

The smell of old leather and horse sweat,

mixed with that of steer manure, greeted them.

"What's on your mind? Must be something important to bring you down here this time of year."

Brown wasted no words. "Hear your place is for sale."

LeDuc laughed, "Sure is. But ain't no way a hard rock miner like you could pay for it. And after what Hoover did to the country, ain't no way it could pay for itself."

"Couldn't we make a deal?" Brown asked.

"Not likely. I want out clean. No down payment and have it kick back with everything run down. Everything but the scrub stock sold off, and taxes three years delinquent."

"But suppose I come up with mostly cash?"

LeDuc looked at him long and hard. He knew what was coming. But suppose Lester had inherited some money from an unknown uncle? If he should suggest that the money came from illegal gold, the story that he had accused a good honest citizen of highgrading would not sit well with the Cousin Jacks of Grass Valley and Nevada City. He decided that he best approach the subject from the uncle hypothesis.

"What happened? You have a rich uncle die?"

Brown said "No, but my credit is pretty good with the banks."

"Oh, hell. You got to go to the banks, forget it. Ain't no banker in his right mind would put up a nickel. No way you could pay off the loan."

Brown only smiled. Then asked, "Your folks used to work for the Empire?"

"Yes. But what's that got to do with it?"

Brown decided to lay all the cards on the table.

"Oh, hell, Bill. You know God damned well the money to buy this ranch came off the copper plates at the Empire. Let's talk turkey."

"I got the money, but I can't give you more'n a token down payment from the sale of my house up on Richardson Street."

He reached into the pocket of his denim jacket, the one which sagged suspiciously from something very heavy. "Here's proof. But I can't give it to you. Deed would have to go through the Title Company. Old MacBoyle would know within hours. I'd go to jail, and you'd lose the best deal you ever made."

LeDuc nodded an acknowledgement when he saw the bag of flattened free gold from the ball mill. "What you got in mind?"

Brown grinned. He saw the gleam in the rancher's eyes. The deal was already made.

"Nothing like real hard gold to tempt a seller," he thought.

"Well, I pay you what I get for the house to bind the deal. Then you get twenty percent of the balance each year until it's all paid. You been asking fifty thousand for the total layout. I'll give it to you. What do you say?"

LeDuc grabbed the offer. "I say yes. But I have to be sure Mary will sign for the deal. Not legal without that. And you have to put the gold in a safety deposit box where we both have to be there when it's opened.

"You don't have to draw me any pictures as to where the gold came from. But why do you want to sink it into these brown brush hills?"

"Well, you know I spent a year in Weimar. Then three more years working as a God damned gardner before MacBoyle let me go underground again. With my lungs, that can't last, and I can't live on this gold in retirement. Everybody would know. Anyway, I don't want my two boys going into the mines and getting what I got. They might not shake it like I did. Here we can all work out in the open. We'll make it give us a living, one way or another."

LeDuc grinned, "Unless something happens to change the price of beef, you'll play hell making any money out of cattle raising."

Brown only nodded, "That's my problem. Tomorrow morning at ten, at the title company?" he asked.

LeDuc stuck out his hand. "See you," he said, and watched Brown drive away in the pickup that, without his knowledge, played such an important part in the sale of his ranch.

In 1942, the Army condemned all the land between the Sacramento Valley and the Indian Springs Road. Camp Beale was built. Tank Divisions were trained for the invasion of North Africa. The semi-desert, the dry and rocky hillsides so closely resembled the terrain where Rummel and his forces ruled the shores of the Mediterranean Sea, that no better place could be found.

Brown did better than most. The title company verified that he had paid fifty dollars an acre for the land. His neighbors got half as much. The War was on. Prices had already started to escalate. He sold the white-faced cattle for double the amount he paid for them.

When the War was won, the newly organized Air Corps needed only the flat and nearly flat lands for the Beale Air Force Base. It sold the remainder.

Normally, when the government sold land which had only recently been obtained by condemnation, the former owners were given the opportunity to buy it back for the sum they had received. In this case, however, the "eager beavers" of the General Services Administration saw a chance to prove to the public, that it was looking after the welfare of the people. The 10,000 acres of surplus land were sold to the general public for from one hundred to one thousand dollars per acre. The city of Marysville was already spreading towards the Air Base land.

Brown only rocked a little more gently on his front porch. He smiled as he thought: "Well, they beat me out of my ranch. Makes us about even. I never had to pay any income tax on the gold, nor on the money the government paid for my land."

Idaho-Maryland Mine, Grass Valley, Calif.

— Photo Courtesy of Nevada County Historical Society

CHAPTER XXIX

THE LAST HIGHGRADE

The deep quartz mines closed in the mid 1950's. Rising costs stopped any thought of reopening the hydraulic placer mines. With few minor exceptions, organized gold mining ceased in California.

But strangely, highgrade appeared from time to time. The Postscript of this book tells of two ex-miners attempting to smuggle stolen gold into Mexico. That was almost ten years after the closing of the mine they took it from. This, the final chapter will tell of the last large cache of highgrade to come to light. This was in 1971. It was no longer considered illegal. The owner of the land on which it was found could, and did, dispose of a part of it at the Mint in San Francisco.

The story is long, and somewhat involved. But if the reader has come this far, he will be intrigued with the story of Bill and Marian Ghidotti.

In 1925, the two graduated from high school. Within a month, they were married.

Bill was a big, bumbling, Italian boy whose grandparents migrated to the mining towns of California, straight from a farming district of Italy. They opened a boarding house for single miners in Nevada City. In time, that graduated into a small hotel.

Bill was not interested in catering to the public. He had his eyes on that new two-ton Mack truck that the P.G. & E. Company had just sent to Grass Valley to replace the five teams and wagons which previously hauled materials to company building and repair sites.

Dan Stewart, the local superintendent, looked over Bill's application. Of course, he already knew Bill. His school record showed that he could at least read and write. He was big, he had a wife to support. He probably would be willing to follow orders and was anxious enough to keep his job to refrain from looking for a shady spot to sleep away the long hot afternoons. Steward gave him a trial.

His first job was to dig holes in the hard-baked red clay to set power poles. He worked hard, and was soon promoted to be an apprentice lineman.

That was not the direction in which he wanted to be advanced. He still had his eye on that truck. He also had learned how to work his way within the organization. His fingers suddenly became all thumbs. He could never learn to use the climbing irons correctly. After his second fall, Stewart took him off the lineman crew. He gave Bill every dirty job in the system to test his staying power.

Bill did the jobs without complaint. Eventually, he was given the job as helper on the truck. At the end of two years, a second truck was sent up from Sacramento and Bill moved behind the wheel of the old Mack. His boyhood dreams had come true. Trucks were still new in the mountain towns. He became the hero of the small boys, replacing the engineer of the Nevada County Narrow Gauge Railroad.

Meanwhile, Marian, who was known as Marian Thomas [after Saint Thomas, a last name given to her by the nuns at St. Mary's Convent, because she was delivered to them from a garbage can at the ripe old age of not more than ten hours] put her training in the commercial school at the convent to good use.

She got a job in the County offices. That

lasted until the Depression came along in 1929. Meanwhile, she and Bill were a few dollars ahead. They made a down payment on the run-down house they were renting. With prices of building materials falling like rocks off Glacier Point in Yosemite and with Bill's job secure because the mines were already expanding, they began to remodel the house.

Given the chance, Marian turned out to be a woman who knew what other women wanted in housing. Shortly, she also proved herself to be a near genius in handling money. The net profit was five hundred dollars on the first house.

Immediately, she placed an option on five more houses, moved into one, and began to remodel.

Rumors were now circulating that Roosevelt was about to double the price of gold. The deep quartz mines exploded with activity. The second house netted two thousand dollars. She ran that two thousand into a quarter of a million before the War intervened and stopped the influx into the mining towns.

Meanwhile, Bill did not relax. She would not let him. He continued to drive a truck for the P. G. & E. Company. Each day after work and each Sunday after early Mass, he did the heavy work on her remodeling projects.

Then they moved into a house with a very large rear yard. Bill was allowed to drive his truck home at the end of each day. He always managed to have the truck loaded with discarded power poles at quitting time. Of course, he unloaded the good dried wood in his own back yard. In time he accumulated enough to start a firewood business on the side.

Marian saw the great demand during the war years for construction and the building of war-time housing. She reasoned correctly that many of the war workers would choose to remain in California. They would be joined by their friends from the East, and from the South; and by the soldiers and sailors who had been stationed on the Pacific Coast or had passed through on their way to, or on their way back from, the Pacific war zones.

On either side of Highway 20 on Washington Ridge above Nevada City, were thousands of acres of timber land which had been cut over 30 to 50 years before. In that area, ponderosa pine and Douglass fir trees make a remarkable growth of nearly a half inch in diameter each year. Now the "small stuff" which had not been disturbed, made logs twenty to thirty inches in diameter. They bought ten sections, all of which had only a downhill pull to a mill in Nevada City.

They then bought the old saw mill which had been standing idle all through the Depression. They sold most of the old machinery for junk and equipped the mill with the best and most modern saws, carriages, and planers that were to be had.

Bill quit the P.G. & E. Company and his beloved trucks. He supervised the mill. Marian gave up the remodeling and became the business manager.

They made money from the start. Lumber which sold for $10 to $15 per thousand board feet before the Depression, now brought $30 to $50, at the mill. They made nothing but money.

When the war ended, Marian saw that once the pipeline filled by lumber from the great mills in the Northwest, competition would once again enter the lumber business. She sold the mill and the timber land for a million dollars and began to think about retiring.

They had a small ranch a few miles below Nevada City on Deer Creek. Bill was not particularly interested, but she loved animals. Especially, she loved to ride and to raise good horses. The land which surrounded them was already being split up into five and ten acre home sites. Fences crossed the trails where she had ridden before. She moved to where she could ride free in God's great outdoors.

They found a ranch in the foothills of the Ruby Mountains in Nevada, fifty miles south and east of Winnamucca. The "home ranch" was not large, but there were 10,000 acres of open range land surrounding them.

Marian was not through placing her furniture when a man from southern California dropped by. He wanted to buy her ranch.

She told him that it was not for sale.

His reply was: "Well, my old daddy always told me that almost everything could be had if the price was right. I know from the county records that you paid a hundred thousand for this place. I'm ready to give you a twenty-five thousand dollar profit and pay your moving expenses to anywhere you want to go. You have to agree with me that a 25% profit for the use of your money for only a few months is much better than most investments."

"No thanks," she said, "it's still not for sale."

"Thirty thousand?" He asked.

She shook her head.

"Forty?" He offered.

She still said, "No." But he caught a gleam of interest in her eyes.

He rose from the table where she had served him a cup of coffee. He walked to the door. He turned the knob; then hesitated.

"Name your price."

She did not blink. But her lips did compress into a straight line.

"Seventy five," she said without a quiver in her voice.

He rose to the bait. "Fifty. And you can live here until June first of next year."

"Sold," she said, and reached out with a firm

handshake. Then she added, "Mister, in this country, this is better than any contract ever written. And you damned well better be able to deliver. Cash in the bank on January first. You can't put that up, you can back out right now. Otherwise, stay clear of this country."

He grinned broadly. "Mrs. Ghidotti, you drive a hard bargain. I'll go you one better. I'll write you a check right now for the entire amount. It's good, and that makes us even. Your handshake, and my cash. How could any two people make a more binding deal?"

"Just one thing more. Suppose your husband refuses to sign the deed?"

A cold hard gleam came into her eyes. "Bill's a good man around machinery, and things like that. But I handle the business affairs in this family."

He nodded, but said nothing.

She smiled and motioned him to return to the table. "I think a deal like this calls for something stronger than coffee. Name your poison, as they still say in this country. Then tell me why you were so all-fired hot on getting this particular ranch. Find a gold mine out back of the corrals?"

"No," he said. "No gold mine. Something better for me. I've enjoyed doing business with you. Never saw a woman quite like you before. But I still have to be sure. Let me write that check. You take it and I'll tell you."

A look of anger crossed her face. "You don't trust me?"

"Sure I do. But I know only a few men I would trust in a deal like this. Just a little precaution."

He handed her the check.

"Three days ago, I found that Bing Crosby bought the ranch next to yours, to the south. He's got a bunch of boys. Little wild sometimes, but he'll tame them. I've got three girls about their ages. Thought it might do some good if the bunch of them got together up here on the long summer evenings. Never can tell what might happen."

She grinned at him. "Well, I'll be damned! A hundred and fifty thousand match-making investment. Biggest gamble I ever heard of."

He only laughed. "Not so big. Maybe ten percent of my next picture. And anyway, as long as Bing owns that property down there, I can always get my money back from some star struck sucker.

It was her turn to laugh. "You suckered me before I had a chance to find out about Crosby."

"Well," he said, "you win some, and you lose some."

So, the Ghidottis returned to Nevada County. They deposited the check, and began to look for another investment. As usual, everything they touched turned to gold. Within the year, they found one which made the fifty percent profit on the ranch look like a month-old bag of potato chips.

Earl MacBoyle of the Idaho-Maryland Mine died in 1948. His widow had no idea of how to manage money or property. The mine closed, but not before she ran up a large deficit. She searched for help.

What she found was a handsome, articulate, stock and bond salesman. He saw in her, a wealthy widow. A "plump goose" just waiting to be plucked. He married her and took over her affairs.

During the following three years, he sold all the MacBoyle liquid assets. He borrowed heavily from the banks, pledging the Idaho-Maryland Mine and all its holdings as security. Then he failed to make any of the payments as they came due.

The banks foreclosed. All the mineral

rights, all the buildings, all the unsold machinery, and all the 1,200 acres of the mine owned land; all of it lying within two miles of the city center of Grass Valley, went under the hammer at a sheriff's sale. The minimum bid was set at $50,000.

Marian Ghidotti drew the $50,000.00 from the Nevada Ranch Sale. She gave Bill the cash and told him to flash the roll of bills as he made the first bid. If anyone wished to make a higher bid, they did not admit it. Bill could bid higher than anyone else who knew the value of the property being sold. Just as Marian had reasoned, they got the Idaho-Maryland Mine.

Did they get a BUY? Let the reader decide. Within three years the government removed the restrictions on the price of gold. The owner could seek the highest market. Immediately, Bill was offered a million dollars in cash for the mineral rights alone. Marian would not let him sell.

At the same time, the value of good subdivision land close to Grass Valley and Nevada City climbed to a minimum of $5,000.00 per acre.

But this is the story of gold and the highgraders.

The State Highway Department decided that a "freeway" must be built through Grass Valley and on through Nevada City. Midas touched the Ghidottis on the shoulder once again. Freeways call for great amounts of concrete. Concrete requires gravel or crushed rock. Good gravel was ten miles away on Greenhorn Creek. Excellent, already broken, hard rock lay in great heaps in the waste dumps of the closed mines. Bill sold the waste dump at the Brunswick Mine. 200,000 tons at 15 cents per ton. (Later the 500,000 tons at the Idaho brought 20 cents.)

The aggregate company placed a crusher by the Brunswick dump. A large power shovel fed the rock into the rock breaker.

To be sure that nothing was disturbed other than the rock, Bill hired a 24-hour watchman to live on the property.

At the end of two weeks of crushing, the watchman came to Bill on Sunday evening. He showed Bill several pieces of quartz shot full of gold. He told the story of how he found it. The story would have been unbelievable anywhere but in the Nevada County gold country.

During the afternoon before the watchman's visit, a man drove up to the Brunswick dump in a fancy new car. He introduced himself and said that before the mines closed, he had worked in that mine. Today, he had brought his family up from Sacramento to show the kids where he had worked underground.

He noticed right away that the power shovel was working in a location of the dump where he had a personal interest, because a year before the mine closed, he was working on a stoping crew on the 3,000 foot level. They shot down a big bunch of highgrade. They had been told that the mine was to close the following month. They figured that the gold would only go to pay off some of the mine's bank debts. It would be a shame to let that happen.

They filled a half dozen powder boxes with the highgrade and held it until the ore bin at the shaft was empty, then marked the next car to be loaded with waste rock. One of the miners suddenly got sick and had to ride the car to the surface. He watched to see where that car was dumped, then took cross bearings and gave each of the other three a crudely drawn map.

The ex-miner told the watchman that according to the map, he knew that the power shovel was getting close to where their treasure was buried. He couldn't do anything about it. The watchman might just as well pick up an extra few month's salary.

He did better than that. He knew that Bill would give him a "cut" of what might be recovered. So, here he was, suggesting that the next 5,000 tons of the rock be put over a sorting belt on its way to the crusher.

When asked if he got all of the $50,000.00 they had paid for the mining property, Bill only smiled. To his friends, he said, "I can't talk about things like that. Marian has to write a big enough check to the IRS as things stand now."

In a way, he was right. But he did not send the gold to the Mint. He kept all those beautiful pieces of rock to show to prospective buyers, if and when he decided to sell the mineral rights.[1]

When the gold reaches the Mint, (if it ever does) the last of the highgrade in large amounts will have left the Gold Country of California.

Hereafter, the gold flowing from the mining country will be in the form of currency brought to the hills by the people anxious to escape the cities. They have discovered that the country lying between 1,000 and 3,000 feet above sea level, the elevation where most of the gold was found, and the elevation lying above the smog and the fog of the Great Valleys and below the snows of the high country, is ideal for retirement living. It is also closer to more and to better recreational activities than almost any other place in the United States.

It has been more than 140 years since the

---

[1]In researching for the stories of the highgraders, the writer asked Bill if he knew of any good ones. He said, "Come with me." He took me to his study and told me to pick up the two-quart fruit jar. I did. But it was so heavy that I had to use two hands. He need not draw me a picture. I knew what was there. So he told me the story of the Brunswick gold.

first gold was taken from the hills of California. During the next 140 years, the amount of gold brought back to the hills by retired couples and by men and women who work in the Central Valleys but commute to the hospitality of the hills, will far exceed the value of all the treasure taken by all the miners since James Marshall picked up that first piece of yellow metal from the mill race at Coloma.[2]

---

[2]Bill Ghidotti died in the early 1970's. When the estate was settled, Marian with her one-half of the community property, plus all but a small amount he willed to some of his nieces and nephews in Italy, now had sufficient cash and other assets to become the principal stock holder (more than 50%) in the GOLD COUNTRY BANK in Glenwood. [Now the West American Bank.]

Upon Marian's death in the early 1980's, a few charities received some cash. The remainder was placed in a trust fund to send deserving students from the Nevada Union High School, through college.

Currently, one boy and one girl from each graduating class of the Nevada Union High School is guaranteed an education in his or her chosen field. At last report, there were 10 students in either a medical school or in a school of veterinary medicine, at the University of California at Davis. Another 10 are in four or five year courses. All their expenses are paid.

POSTSCRIPT

In 1964, Mike McGuire, an old-time "hard rock miner" died in Berkeley, California. We took the body back to Grass Valley to be interred in the family plot.

While there, one of Mike's nephews, a mining engineer, asked if I knew that two "highgraders" were caught as they tried to carry gold into Mexico where it could be turned into Mexican then into U.S. currency?

As the mines had been closed for nearly ten years, my first impulse was to ask how they got their hands on the gold? I stopped in mid-sentence. "No," I said, "Don't tell me, I'll tell you."

"They took the gold out of the Empire or the Idaho-Maryland while those mines were still operating. They buried the rich ore or the amalgam in their back yards. They expected to use it as an annuity following their retirement: dig up enough each year to live on comfortably.

What they had not foreseen was that newly mined gold could only be sold to the United States Treasury. Now all the mines were closed because the Government would pay only $35.00 per ounce, the amount fixed by law in 1933. No more gold was being delivered to the U.S. Mint in San Francisco. Their gold was worthless unless it could be sold outside the United States.

The two ex-miners needed eating money. Hence the smuggling attempt which ended in tragedy."

The mining man laughed, then asked how I had been able to solve the mystery so quickly?

My reply was: "That I had not lived with, and been associated with mine workers for more than thirty years, without learning something about the mines and the miners."

Gold and the Highgraders

During the drive back to Berkeley, I began thinking of the treasure trove of stories I knew about highgrade and the highgraders.

If someone did not consign them to the printed page, these stories were about to be lost to future generations. As I drove, I made a mental survey of the dozen or so men who knew as much, or more, than did I about the activities of the miners who, by one means or another, took gold-studded quartz rock from the deep mines or gold amalgam from the stamp mills.

Only the mine superintendents or a few other management personnel had the education needed to write the intriguing stories. This, they could not do. Any attempt on their part to publish the stories, would associate them with the criminals. "If they knew what was going on, why had they not turned the thieves in to the authorities?" They were compelled to remain silent.

Except for scientific papers, I had never written for publication, but I was determined to write a few of the stories as I knew them. At least my grandchildren would have a record of the activities of their relatives and of their friends who still lived in the gold mining districts of California.

In the beginning I anticipated a long magazine article or a very small booklet. Once I became engrossed in the task, the typewritten pages multiplied until I had the manuscript for a book.

I took the material to the West Coast Editor of one of the largest publishing companies in the world.

Within the week, the editor called me at my home. "Your Highgrader stories are wonderful. We want to publish!" he said.

For the next several weeks, I did not walk on "cloud number nine," I jumped all the way to cloud ninety-nine.

Eventually, the great company decided not to publish.

With great confidence I took the rejected manuscript to the Howell-North Publishing Co., in Berkeley. At that time, this was the largest and best-known publishing company on the west coast. Mr. North was a retired mining engineer. He had once worked in the mines I had written about. He read the manuscript and accepted it for publication.

Among other things, he said: "This is a major contribution to the history of mining, and to the history of California as well."

Later, he added: "I don't know how anyone other than a mining engineer or a mine superintendent could have written such a knowledgeable and interesting book. But then, the professional men, had they tried, would have become so involved in details and in technical material, that they would have lost their readers."

Eventually, Mr. North and I disagreed about using a story of Herbert Hoover, another ex-mining engineer, and former President of the United States. North claimed that Hoover would never have carried on the activities I told of. I insisted that the story be included. I had been on the spot when it happened and I had documents to prove my points. (The Scott Bar Incident.)

I took the manuscript to a third, and then to a fourth publishing house. Each accepted the book on first reading. Then each, in turn, decided against their original decision.

Upon careful inquiry, I eventually discovered that the publishing companies had submitted the manuscript to their attorneys. Each lawyer in turn advised against publication for fear of libel.

In the writing, I changed all names—except those of historical characters, such as Hoover. I changed the location of most of the mines I named, but the stories in some cases were so vivid that someone was bound to recognize "Old Muley Jim." His heirs were sure to bring suit for libel and/or for defamation of character.

During the ensuing years, the activities of the Highgraders have come to be looked upon as those of modern Robin Hoods. "Taking from the rich to give to the poor." Today, referring to someone as a "Highgrader," in the mining country would be laughed out of court if indeed legal action were taken.

After the fourth rejection, the manuscript was put aside. It lay on a shelf gathering dust until now.

Meanwhile, I have written four other books, each based on the mines and the miners.

In order of publication, they are:

1. A Self Guided Tour of the Grass Valley - Nevada City, Mine Sites. (Including detailed maps.)
2. 49er Irish. One Irish Family in the California Gold Mines.
3. Coolies, Kanakas, and Cousin Jacks. 14 Ethnic Groups, in all, are the basis of stories of the great migration of the strangers into California and their contributions to the mining industry and to the making of a great state.
4. The Lassen Trail. The true story of an 18 year old boy who came via that route in 1849.

California Gold and the Highgraders completes the series I had in mind when I began writing of the mines and of the miners.

My purpose has been to bring to the reading public the romance of the founding and growth of a great and growing economy on the shores of the Western Sea.

Sacramento, California
January, 1988

OFFICE, YUBA WATER CO.
OTT'S ASSAY OFFICE (RIGHT)

G L O S S A R Y

ALLUVIUM: Material deposited by running water.

AMALGAM: One metal dissolved in another — e.g., gold in mercury.

APEX: The point where the ledge contacts the surface; an outcropping.

ARGONAUT: A forty-niner (from the searchers for the Golden Fleece in Greek Mythology).

ARRASTRE: A Mexican ore grinder - a large rock is dragged over the ore to pulverize it.

ASSAY: To test ore for the value of the metal it contains.

BALL MILL: A revolving steel drum partly filled with steel balls which crush the ore as they roll over and over.

BAR: Gravel left at water level by the shifting of the stream channel.

BEDROCK: Hard, uneroded stone on a stream bottom or under soil.

BONANZA: A very rich strike.

BULLION: Gold melted and cast into bars.

CHANNEL GOLD: Nuggets and dust in the river and stream gravel.

CHIMNEY: The vertical section of a quartz ledge containing gold.

CLAIM: A section of ground given by the government for mining purposes.

G L O S S A R Y

CLASSIFIER: A device for separating crushed ore into various sizes.

CLEAN-UP: Gathering the amalgam from the sluice-boxes; the final step in the milling process.

CONCENTRATES: The collection of valuable minerals extracted from the ore.

CONTACT LEDGE: Quartz lying between two types of country rock, i.e., between granite and serpentine.

COYOTING: Prospecting by digging straight down in the hope of intercepting a paying ledge or gravel channel.

CROSS-CUT: To tunnel at right angles from a drift or ledge.

CRUSHER: A rock breaker.

CYANIDE: A compound of carbon and nitrogen; in mining, either potassium or sodium cyanide (KCN or NaCN).

DEBRIS: Waste material; material washed out by the hydraulic mines.

DIAMOND DRILL: A drill consisting of iron pipe whose end is studded with industrial diamonds; used to core and withdraw samples of rock for assay.

DIGGINGS: The location of mining activity.

DIP: The angle the ledge makes with the horizontal.

GLOSSARY

DISCOVERY HOLE: A shaft sunk on mining property before a claim becomes legal.

DREDGER: A mechanical device for digging gravel and separating it from any gold it may contain (usually a very small amount).

DRIFT: A horizontal tunnel, usually following a ledge.

DUST: Very small particles of gold.

EOCENE EPOCH: A geological age (part of the Tertiary).

ERODE: To wear away, usually by water.

FISSURE: A weak area in the mantle rock of the earth's crust; a crack.

FLOTATION: The process of separating metallic minerals from pulverized ore.

FLOTATION CELL: A device for carrying out the flotation process.

FLUME: A trough to carry water; an aqueduct.

FOOL'S GOLD: Iron pyrite, iron disulfide (FeS_2), so called because it resembles gold in color.

GROUND SLUICE: To wash away gravel and pick up the gold trapped behind the irregularities in the bedrock.

HANGING WALL: Rock above a ledge; ceiling.

G L O S S A R Y

HIGHGRADE: Any rich rock, usually quartz with visible free gold; also, to steal such ore from a mine.

HYDRAULIC MINING: Washing gravel from a bank by playing a stream of water under high pressure against it.

JACKHAMMER: A rock drill operated by compressed air.

KINETIC ENERGY: The energy possessed by a body in motion.

LEDGE: A layer of quartz rock between ordinary rock.

LEVELS: The working stages in a mine, usually at 100-foot intervals.

LODE: A mineral-bearing ledge.

MESH: The size of the opening in a screen.

MILL ROCK: Ore containing enough value to warrant sending it through the recovery process.

MONITOR: A device for throwing high-pressure water in hydraulic mining; a "little giant."

NUGGET: A chunk of gold.

ORE: Any rock containing a valuable metal.

OUTCROPPING: A ledge coming to the surface of the ground.

GLOSSARY

PAN: A shallow iron dish with sloping sides used for the separation of gold from sand and gravel; also, to wash gravel with such a pan.

PLACER MINING: The extraction of free gold from sand and gravel.

POCKET: A small area highly enriched with mineral.

PROSPECT: To look for minerals; also, a place where minerals may be found, or a place where a miner is searching for minerals.

QUARTZ: The mineral composed of silicon and oxygen (SiO_2), gold usually being associated with it.

RETORT: A device for heating a metal; in gold mining, for vaporizing mercury from amalgam.

RIFFLE: An obstruction interrupting the flow of water to allow heavy material to be deposited.

ROCKER: A device for separating gold from gravel.

SHAFT: A well or vertical opening through which a mine is operated.

SHOOT: A section of ledge containing valuable mineral; a "chimney."

SKIP: A cage or car for the lowering and raising of men and materials in the shaft of a mine.

G L O S S A R Y

SLICKINGS: Very fine tailings from a hydraulic mine.

SLUICE BOX: A flume or box with cleats across the bottom to trap heavy material: gold.

SLUICING: Washing gravel through a sluice box.

SLURRY: The mixture of water and crushed ore to be treated for the separation of its mineral content.

SPONGE GOLD: The porous gold remaining after the mercury has been evaporated from amalgam.

STAKE: To post notices of ownership around a claim; also, money advanced to a prospector to buy food and supplies (grub stake).

STAMP: A heavy weight for crushing ore.

STAMP MILL: The mechanical device for operating stamps.

STOPE: The area where ore is taken from a ledge.

STRINGER: A very thin or narrow ledge.

STRATA: Layers of rock.

TAILINGS: Waste material from a mine; ore, after values have been extracted.

TERTIARY: The second most recent geological age.

TUNNEL: A passage through rock; a drift.

TRIBUTE: A system of employment in the mines whereby the miners furnished the labor and explosives and the mine supplied everything else. The gold was divided equally.

VEIN: A layer of mineral-bearing rock (usually quartz); a ledge.

WEATHERING: The softening of rock due to the action of water, air, and chemicals of any kind.

TWO STORY OUTHOUSE. Located immediately to the rear of the National Hotel. Nevada City. Saved from the bulldozer in 1971, by concerned citizens